“I have used this method to [illegible] facts for three years, and I believe! It gives the child who learns differently an anchor for their learning. They become confident and use the songs to remember their facts. Eventually, the songs do the job and the learning becomes mental. On a recent standardized test I was amused to see the number of students singing to themselves (in their heads) when they were checking their answers. The kids *want* to sing the multiplication songs. It's a fun, meaningful way to learn.”

~Eunice Tritt,
Elementary School Teacher of 25 Years

“Yesterday, my son was excited because he was the only student in his class working on his 11's. That means he's mastered the 0–10's! We really feel [your] workshop was worth it! It has paid for itself every week this school year!”

~Beth Porter,
Parent and Elementary School Guidance Counselor

“I've had a number of students in my classroom that have been taught multiplication using this method. I'll never forget the look on one child's face when I gave a timed multiplication test. This student, who usually complained about such assignments and inability to complete them, exclaimed, ‘This is easy!’ There's nothing more satisfying to a teacher than seeing a child experience success!”

~Susan Ratcliff,
Elementary School Teacher of 7 Years

THE MULTIPLICATION MIRACLE

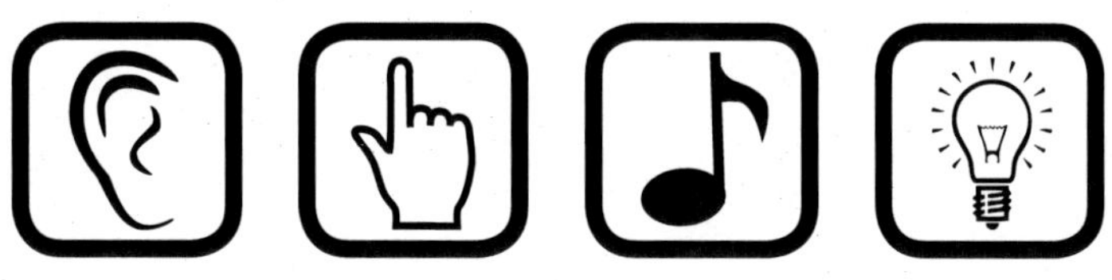

THE MULTIPLICATION MIRACLE

THE HEAR IT, TOUCH IT, SING IT, "GET IT" METHOD

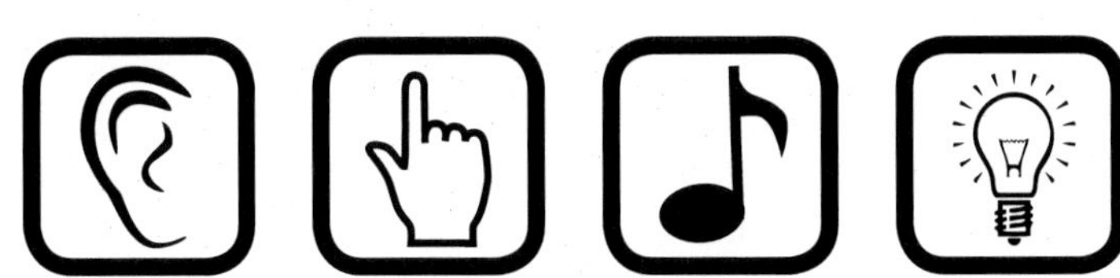

SHANNON J. CARNES

Tate Publishing & *Enterprises*

This title is also available as a Tate Out Loud product. Visit www.tatepublishing.com for more information.

The songs used in this book have been thoroughly researched and are believed to be in the public domain. Proof of this status has been obtained from the offices of the Public Domain Project in the documented form of sheet music with a copyright date of 1922 or earlier. If any of the above is believed to be in error, please contact the publisher immediately.

The opinions expressed by the author are not necessarily those of Tate Publishing, LLC.

Published by Tate Publishing & Enterprises, LLC
127 E. Trade Center Terrace | Mustang, Oklahoma 73064 USA
1.888.361.9473 | www.tatepublishing.com

Tate Publishing is committed to excellence in the publishing industry. The company reflects the philosophy established by the founders, based on Psalm 68:11,
"The Lord gave the word and great was the company of those who published it."

Cover design & layout design by Elizabeth A. Mason

Published in the United States of America

ISBN: 978-1-60462-506-6
1. Education: Mathematics 2. Teaching Methods & Materials
07.11.27

TABLE OF CONTENTS

I. IMPORTANT INFORMATION

II. INSTRUCTOR'S GUIDE: TEACHING THE METHOD

III. STUDENT WORKSHEETS

APPENDIX/FORMS

DISCLAIMER

IMPORTANT INFORMATION

Note to instructor:

It is extremely important that you read the pages in this chapter carefully. I know you're eager to get started, but don't skip over this part. It will be essential in your success with the program.

Dear Teacher,

First, I would like to congratulate you on choosing *The Multiplication Miracle*. I feel certain you will find it as helpful as I did in teaching my students. The first section is considered to be the "Important Information." In this section, Part I of the guide, you will find an overview of the program (a sort of explanation of how it all works) as well as the "Introduction to Songs" and the "Multiplication Songs List" pages (helpful with the singing portion of the method.)

Part II of the book is the "Instructor's Guide: Teaching the Method." This will include an outline of how the chapters are presented and will give you a better understanding of what to expect in each chapter. You will also receive detailed instructions on how to teach the method and deliver instruction on a chapter-by-chapter basis. These pages (p. 11-99) are protected by copyright and may not be reproduced.

The section of the book entitled "Student Worksheets" begins on page 100. This contains reproducible worksheets that will be completed by the learner. These should be given in the order presented in the book. Limited reproduction of these pages is permissible. In other words, reproduce for the students in your class, or the children in your home. Do not, however, make copies for the teacher's class next door, the entire school, neighborhood, etc. Obviously, there will be no hidden cameras watching you, but please realize my time and effort were put into the production of this program and it would be dishonest to disregard copyright law. I have made every effort to make *The Multiplication Miracle* both affective and affordable.

I thank you for your purchase of the program and am confident that you will soon see evidence of the "miracle." I look forward to hearing your success stories.

Sincerely,
Shannon J. Carnes

PROGRAM OVERVIEW

This program is not intended to replace textbook instruction and/or good, old-fashioned memorization of multiplication facts. Students capable of memorization have no reason to be taught using *The Multiplication Miracle*. This program is for those who struggle with concepts requiring memorization. *The Multiplication Miracle* is intended to be a supplementary teaching tool, as well as an alternative to traditional teaching methods.

In teaching students with special learning needs, I found that many learners are simply unable to memorize material. This program does not focus on memorization; instead, it teaches a method—one that the learner can depend on to work every time. The method is fun and simple to learn and use. The method can even be applied to division with only a few modifications.

Learners who traditionally learn at a slower pace, as well as those who are unable to memorize, "get it" with *The Multiplication Miracle*. They retain it and truly understand it, as well. Those who tend to be quick learners are able to catch on to the method immediately and are able to develop a greater understanding of the process of multiplication. Not only that, but most learners are generally able to recall multiplication facts within seconds after repeated practice using this method.

The Multiplication Miracle starts with the basics, making certain that the learner understands what multiplication is—repeated addition or "counting by" a number. Students may have heard this concept explained as "skip counting." Next, a series of fun songs are learned, along with an easy to teach and learn finger-numbering technique.

You will use the "Instructor's Guide" to deliver and guide instruction. You will make copies of the "Student Worksheets" for student use. It may be helpful to make copies of one chapter at a time and either staple or place in a folder for each student. Each chapter in the "Instructor's Guide" corresponds with a chapter in the "Student Worksheets" section. See corresponding chapter titles and references to page numbers found in parentheses, as well as the Table of Contents for assistance. You will find the method easy to teach and the format easy to follow; in fact, you probably won't need the Instructor's Guide for much of the instruction.

On the following pages (15-18) you will find the "Introduction to Songs" and "Songs List" respectively. These will be helpful in teaching the songs. Here they are presented in a way that helps you with the tunes. (Also visit www.getitmethod.com for supplementary teaching materials.) There is no need to be musically inclined to learn or teach these songs.

Included in the Appendix is a generic certificate page and score sheet, which you will also copy as needed. An answer key for the "Student Worksheets" section is also found here.

INTRODUCTION TO SONGS

The "Multiplication Songs" are an extremely important part of *The Multiplication Miracle* method. For Chapters 3 through 6, as well as 8 through 12 and 14, a song will be taught/learned. These songs are used to help the student "count by" a number more easily. There will be songs for the Two's, Three's, Four's, Five's, Six's, Seven's, Eight's, Nine's, Ten's, and Twelve's. Special rules apply to the Zero's, One's, and Eleven's, Therefore, no songs are required.

The "Multiplication Songs" are to the tune of familiar children's songs (found in the public domain.) Most learners have heard these songs at some point, even if not familiar with the song titles; for example, "Camptown Races." Generally, upon hearing the tune, they recall the song. If by chance the tune is not familiar, learners tend to "catch on" very quickly.

The songs will be taught separately in each chapter mentioned above on the page entitled "Learning the Song." In the beginning, focus the learner on singing each word (number) in the song deliberately and slowly. The songs can be sped up to a more natural pace later. Encourage the learner to picture the numbers in his/her mind while singing. Detailed instructions on teaching the songs to the learner are found in the "Instructor's Guide."

Before attempting to teach the songs to the learner, it is essential that you preview the song beforehand. You should study the subsequent pages, "Multiplication Songs" (p. 16 and 18) in depth. First, be sure you know the *correct* tune of the song. If you do not know it, find the song by researching the Internet, a children's song audiotape or album, asking a peer, etc. Once you are familiar with the tune, refer to the "Multiplication Songs" pages. Notice how the numbers for each song correspond with a word or group of words in the song. Practice saying/singing the numbers instead of the original words. There are intentionally no music notes due to the fact that you do not have to "read" music in order to use this method. It is essential that you have the songs perfected yourself before teaching them. Do not worry about the quality of your singing—it is the content that's important!

MULTIPLICATION SONGS LIST

<table>
<tr><td rowspan="8">Two's Song Tune: "The Farmer in the Dell"</td><td>The</td><td>farmer</td><td>in the</td><td>dell,</td></tr>
<tr><td></td><td>2 4</td><td>and</td><td>6</td></tr>
<tr><td>the</td><td>farmer</td><td>in the</td><td>dell,</td></tr>
<tr><td></td><td>8 10</td><td>and</td><td>12</td></tr>
<tr><td colspan="2">high ho</td><td>the</td><td>dairy-O</td></tr>
<tr><td colspan="2">--14--</td><td></td><td>--16--</td></tr>
<tr><td colspan="2">the farmer</td><td>in the</td><td>dell</td></tr>
<tr><td colspan="2">--18--</td><td>and</td><td>20</td></tr>
</table>

<table>
<tr><td rowspan="8">Three's Song Tune: "Jingle Bells"</td><td>Jingle</td><td>bells,</td><td>jingle</td><td>bells,</td></tr>
<tr><td>3 6</td><td>9</td><td>12</td><td>15</td></tr>
<tr><td colspan="2">jingle</td><td colspan="2">all the way</td></tr>
<tr><td colspan="2">18</td><td colspan="2">--21--</td></tr>
<tr><td colspan="2">Oh what fun</td><td colspan="2">it is to ride</td></tr>
<tr><td colspan="2">--24--</td><td colspan="2">--27--</td></tr>
<tr><td colspan="2">in a one horse</td><td colspan="2">open sleigh</td></tr>
<tr><td colspan="2">Don't forget</td><td colspan="2">Old --30--</td></tr>
</table>

Four's Song Tune: "Clementine"	Oh	my	darlin',	oh my	darlin',
	4	8	12	--16--	--20--
	oh my darlin',			Clementine,	
	--24--			--28--	
	you are lost		and		gone forever.
	--32--				--36--
	Dreadful sorry,			Clementine	
	--Finally--			--40--	

Five's Song Tune: "Pop Goes the Weasel"	Round	and	round	the	cobbler's bench
	5		10		--15--
	the monkey			chased the weasel.	
	--20--			----25----	
	The monkey		Thought 'twas		all in fun,
	--30--		--35--		--40--
	pop goes		the		weasel!
	45		And		50

Six's Song Tune: "Camptown Races"	Camptown	ladies	sing this song	doo dah,	doo dah.
	6	12	18	24	30
	The	Camptown	racetrack		five miles long
		36	42		--48--
	oh, doo-dah			day	
	54, and			60	

Seven's Song Tune: "Twinkle Twinkle Little Star"	Twinkle	twinkle	little star,		how I wonder	what you are.	
	7	14	—21—		—28—	—35—	
	Up	above			the world	so high,	
	42	49			—56—	—63—	
	like a diamond	in			the	sky	
	and 70	Can't			you	see?	
	Twinkle	twinkle	little	star	how I wonder	what you	are
	Sevens	are so	fun to	do.	Don't you	think so	too?

Eight's Song Tune: "Mary Had a Little Lamb"				
	Mary	had a	little lamb	
	8	16	--24--	
	little lamb,	little lamb,		
	--32--	--40--		
	Mary	had a	little lamb,	
	48	56	64	
	it's	fleece was white	as	snow.
		--72--	and	80

Nine's Song Tune: "For He's a Jolly Good Fellow"			
	For	he's	a jolly good fellow,
	9	18	27
	for	he's	a jolly good fellow,
	36	45	54
	for	he's	a jolly good fellow,
	63	72	81
	which	nobody can	deny.
	and	then comes	90

Ten's Song Tune: "Three Blind Mice"						
	Three	blind	mice,	three	blind	mice,
	10	20	30	40	50	60
	see	how they	run,			
	70	--80--	90			
	see	how they	run.			
		and	100			
	Three	blind	mice,	three	blind	mice
	Sing	the	tens.	That's	the	end.

Twelve's Song Tune: "Billy Boy"							
	Oh,	where	have you been,	Billy boy,	Billy boy?		
		12	--24--	--36--	--48--		
	Oh,	where have	you been,	charming Billy?			
		--60--	and	--72--			
	I have been	to seek a wife,	she's the	idol	of	my	life.
	--84--	--96--	Then	it's	one	O	eight
	She's a young thing	and	cannot leave	her mother.			
	And last comes		one hundred	--twenty--			

INSTRUCTOR'S GUIDE:

TEACHING THE METHOD

SUMMARY OF INSTRUCTOR'S GUIDE

Each chapter (with the exception of Chapters 1, 2, 7, and 13) follows a similar format and includes the following:

Instructional Pages-

Introductory and primary instruction is given on these pages. You will do a lot of talking and asking questions, as well as guiding the learner through examples. Allow the learner to complete these pages, but be sure you are offering plenty of guidance and opportunity for practice of the concept.

Practice Pages-

The learner will complete these pages individually; however, the instructor will offer guidance as needed, closely monitoring depending on how well you feel the learner understands the material.

Activity Pages-

These pages are to be made into a game or activity for practice. You will distribute them along with necessary materials and give instructions for completion.

Practice Quiz Page-

The learner completes these pages with no guidance from the instructor. The instructor then grades the paper by circling or marking a dot by incorrect answers (refrain from using X marks as this may lead to frustration.) The learner then reworks missed problems with guidance from the instructor.

Quiz Page-

It may be beneficial to complete this page on a separate sheet of paper to allow the learner to retake the quiz multiple times. The learner completes this page with no guidance. The instructor grades the quiz and the learner records it on the Score Sheet (p. 242.)

Certificate Page-

This page is an award certificate completed by the instructor. It is found on page 243. You will make multiple copies so that each learner gets one after completion of each chapter. It is not necessary that the learner achieve 100% mastery on the chapter Quiz Page in order to receive the certificate or progress to the next chapter. Though 100% is the goal, 80% should be sufficient in order to continue. Material from previous chapters is revisited continuously throughout the book for the maximum amount of practice.

CHAPTER 1:
COUNTING BY A NUMBER

Note: The format of this chapter is different from most others in this book due to its content. Because most students catch on quickly to "counting by" a number, you will find an instructional page only for each number, one through twelve.

PAGES 104-115: INSTRUCTIONAL PAGES

1. Instruct the learner to notice the first set of pictures at the top of the page (the ones with the numbers under them). Have the learner touch the objects and count them, noticing the numbers written below.

2. Draw a circle around each group of objects and explain to the learner that each circle is a group. (For example, one group of one, two groups of one, three groups of one, etc.) Ask the learner how many circles or groups in all.

3. Have the learner count and write the numbers below the second set of pictures.

4. Tell the learner to draw a circle around each group of objects. Ask the learner how many objects in four circles (groups), two circles (groups), etc.

5. Assist the learner in "counting by" the given number by writing the numbers in the blanks. Explain how to use fingers to count up, if needed.

6. Assist the learner in "counting by" the given number by writing the numbers in the boxes.

Practice Activities

Provide additional practice, if needed, by drawing groups of simple objects and having the learner "count by" the given number. Use the instructional page as your guide. Students also benefit from counting "real world" objects. Examples include groups of people, crayons, stacks of pennies, etc.

CHAPTER 2:
NUMBERING THE FINGERS

Note: The format of this chapter is different from most others in this book due to its content. There is no quiz for this chapter, so there is no space on the score sheet for recording the grade. For additional practice, see the suggestions found on the following page.

PAGE 118: INSTRUCTIONAL PAGE

1. The learner will place his/her hands over the handprints on the page. Explain that anytime the hands are used to multiply, they must be placed palm side down, as pictured. It may be helpful to tell the learner that the hands are placed dirty side down (if hands were messy, the dirt couldn't be seen) or ring side up (if rings were being worn, they could easily be shown off).

2. Explain to the learner that the left pinkie finger is numbered as "finger number one," noticing the "1" in the picture. The right pinkie finger is "finger number ten," noticing the "10" in the picture. Count off the fingers in order.

3. Have the learner practice counting off the fingers, wiggling each finger as it is counted.

4. Point to individual fingers on the learner's hand and ask the learner what "number" each finger is called (ex. left ring finger is finger number 2)

5. Ask the learner to wiggle random fingers (ex. "wiggle finger number three"). Continue until the learner understands the process.

6. Have the learner complete the coloring portion of the page.

PAGE 119: PRACTICE PAGE

1. The learner will draw lines from the number to the corresponding finger.

2. Tell the learner to place his/her hands on the picture and count off the fingers, wiggling each one. Try this again, except touch fingers down lightly on the paper.

3. Have the learner complete the coloring portion of the page, as well as numbering the fingers in the blanks.

4. Point to individual fingers on the learner's hand and ask learner what number each finger is called as previously described.

Additional Practice: Other ideas for extra practice

Have the learner complete the following activities:

- Trace hands on paper. Number the fingers. Color even-numbered fingers green and odd-numbered fingers blue.

- Trace hands on paper. Draw rings on fingers. Decorate the rings with correct numbers.
- Hold hands up in front of body and say "number 3 finger" and bend that finger down. Continue until all fingers "disappear."
- Draw hands on sidewalk with chalk and number them.
- Think up your own creative ideas!

CHAPTER 3:
LEARNING THE TWO'S

PAGES 122-124: INSTRUCTIONAL PAGES

Page 122: "Learning the Song"

1. Read the words of the "Two's Song" aloud without singing the tune. Have the learner read the words or repeat after you. Tell the learner to notice the picture clue. Explain that the tune of this song is "The Farmer in the Dell."

2. Sing the song to the learner. (See page 16 for assistance with song.) Explain that the picture shows a farmer with two pigs. Because of the number of pigs the farmer has (two), the student will be reminded that the "Two's Song" is to the tune of "The Farmer in the Dell." Have the learner color the two pigs red.

3. Have the learner practice singing the song multiple times until it becomes very familiar. The learner should practice the song as much as possible. Consider singing while at recess or quietly during transition times. Also encourage the learner to practice at home, in the car or on the bus, while taking a walk, etc. Continue to remind the learner of the picture clue that accompanies the song.

Page 123: "Numbering the Fingers"

1. Tell the learner to place hands over the handprints on the page. Tell the learner to recall how the fingers are numbered. Review Chapter 2 if needed.

2. With hands on the handprints, have the learner sing the "Two's Song" aloud, remembering the picture clue. The learner should wiggle or touch down each finger in order while singing the song. Be sure to sing each word (number) of the song deliberately, pausing on the correct finger.

3. The learner will write the song in the blanks above the fingers. Remind the student of "counting by" a number, as in Chapter 1, and that the song is the same as "counting by" two's.

4. The learner will complete the coloring portion of the page.

5. Have the learner practice singing the song while touching down the tips of the fingers to the paper. After the learner seems to understand the process, remove the worksheet and have him/her practice the same way, touching down fingers to any surface, such as a table, book, desk, or lap (even the air will work).

Page 124: "Learning to Multiply"

1. Have the learner look at Example A. The problem is 2 x 4 = __________.

2. Explain the following method to the learner: The first number in the problem is 2, so the "Two's Song" will be sung to

find the answer. The second number in the problem is 4, so the fourth finger is where the learner will stop singing. (The learner may imagine a miniature stop sign over the fourth finger, or one can even be made out of a bottle top or scrap of construction paper.) It is very important that the "stop" finger be identified and touched down. The number the learner is singing when he/she gets to the "stop" finger is the answer. In this case, the number being sung when the fourth finger touches down is eight. Therefore, the answer is 8.

3. Follow this procedure to guide the learner through all examples on the page.

PAGES 125-126: PRACTICE PAGES

The learner will complete these pages with assistance from the instructor as needed. The instructor should monitor the learner's work to ensure the learner is using the finger-numbering method with songs and finger movements. Remind the learner that once a certain fact has been memorized, there is no need to continue using the method for that fact.

PAGES 127-128: ACTIVITY PAGES

Page 127 "Count by Cards"

1. Glue the page to a small piece of poster board or other heavy paper.

2. Cut on the lines.

3. Write the number two on the back of each card as a reminder that these cards go with the "Two's Song."

4. Use the cards to play the following games or make up your own. Store cards in a resealable plastic bag or envelope, or secure with a rubber band.

Games:

Turn cards face up on a flat surface and mix up. Have the learner see how quickly he/she can put them in order. A kitchen timer or stopwatch can be used. Another option includes placing cards in order and practicing touching down fingers and singing songs.

Page 128 "Multiplication Cards"

1. Follow steps 1 and 2 from "Count by Cards."

2. Write answers on back of cards and store in resealable plastic bag or envelope, or secure with a rubber band.

3. Use cards to practice using the finger-numbering method and songs to multiply. Then check answers by flipping over cards to the back side. This game can be played individually or with multiple players.

Page 129: Practice Quiz Page

1. The learner completes Quiz A with no guidance from the instructor.

2. The instructor grades the paper by marking incorrect answers.

3. The learner should rework missed problems with guidance from the instructor.

4. Quiz A should be covered up with a folded sheet of paper or a book. Have the learner complete Quiz B.

5. Repeat steps 2 and 3 for Quiz B.

6. Quiz B should be covered up and Quiz C completed.

7. Repeat steps 2 and 3 for Quiz C.

PAGE 130: CHAPTER QUIZ PAGE

It may be beneficial to complete this page on a separate sheet of paper to allow the learner to retake the quiz multiple times.

1. The learner completes this page with no guidance.

2. The instructor grades the quiz and it is recorded on the Score Sheet (p. 242).

3. If score is 80% or above, the certificate page (p. 243) may be completed; however, a lower score may indicate a need for more practice. If more practice is needed, the learner should redo chapter pages as well as spend more time practicing the song and completing game activities.

CHAPTER 4:
LEARNING THE THREE'S

PAGES 132-134: INSTRUCTIONAL PAGES

Page 132: "Learning the Song"

1. Read the words of the song aloud without singing the tune. Have the learner read the words or repeat after you. Tell the learner to notice the picture clue. Explain that the tune of this song is "Jingle Bells."

2. Sing the song to the learner. (See page 16 for assistance with the song.) Explain that the picture shows three bells. Because there are three, the learner will be reminded that the "Three's Song" is to the tune of "Jingle Bells." Have the learner color the three bells red.

3. Have the learner practice singing the song multiple times until it becomes very familiar. The learner should practice the song as much as possible. Consider singing while at recess or quietly during transition times. Also, encourage the learner to practice at home, in the car or on the bus, while taking a walk, etc. Continue to remind the learner of the picture clue that accompanies the song.

Page 133: "Numbering the Fingers"

1. Tell the learner to place hands over the handprints on the page. Tell the learner to recall how the fingers are numbered. Review Chapter 2 if needed.

2. With hands on the handprints, have the learner sing the "Three's Song" aloud, remembering the picture. The learner should wiggle or touch down each finger in order while singing the song. Be sure to sing each word (number) of the song deliberately, pausing on the correct finger.

3. The learner will write the song in the blanks above the fingers. Remind the student of "counting by" a number as in Chapter 1, and that the song is the same as "counting by" threes.

4. The learner will complete the coloring portion of the page.

5. Have the learner practice singing the song while touching down the tips of the fingers to the paper. After the learner seems to understand the process, remove the worksheet and have him/her practice the same way, touching down fingers to any surface, such as a table, book, desk, or lap (even the air will work).

Page 134: "Learning to Multiply"

1. Have the learner look at Example A. The problem is 3 x 5 = __________.

2. Explain the following method to the learner: The first number in the problem is 3, so the "Three's Song" will be sung

to find the answer. The second number in the problem is 5, so the fifth finger is where the learner will stop singing. (The learner may imagine a miniature stop sign over the fifth finger, or one can even be made out of a bottle top or scrap of construction paper.) It is very important that the "stop" finger be identified and touched down. The number the learner is singing when he/she gets to the "stop" finger is the answer. In this case, the number being sung when the fifth finger touches down is fifteen. Therefore, the answer is 15.

3. Follow this procedure to guide the learner through all examples on the page.

PAGES 135-136: PRACTICE PAGES

The learner will complete these pages with assistance from the instructor as needed. The instructor should monitor the learner's work to ensure the learner is using the finger-numbering method with songs and finger movements. Remind the learner that once a certain fact has been memorized, there is no need to continue using the method for that fact.

PAGES 137-138: ACTIVITY PAGES

Page 137: "Count by Cards"

1. Glue the page to a small piece of poster board or other heavy paper.

2. Cut on the lines.

3. Write the number three on the back of each card as a reminder that these cards go with the "Three's Song."

4. Use cards to play the following games or make up your own. Store cards in a resealable plastic bag or envelope, or secure with a rubber band.

Games:

Turn cards face up on a flat surface and mix up. Have the learner see how quickly he/she can put them in order. A kitchen timer or stopwatch can be used. Another option includes placing cards in order and practicing touching down fingers and singing songs.

Page 138: "Multiplication Cards"

1 Follow steps 1 and 2 from "Count by Cards."

2. Write answers on back of cards and store in resealable plastic bag or envelope, or secure with a rubber band.

3. Use cards to practice using the finger-numbering method and songs to multiply. Then check answers by flipping over cards to the back side. This game can be played individually or with multiple players.

PAGE 139: PRACTICE QUIZ PAGE

1. The learner completes Quiz A with no guidance from the instructor.

2. The instructor grades the paper by marking incorrect answers.

3. The learner should rework missed problems with guidance from the instructor.

4. Quiz A should be covered up with a folded sheet of paper or a book. Have the learner complete Quiz B. Be sure to point out that Quiz B contains numbers multiplied by two, thus a review of previously learned material.

5. Repeat steps 2 and 3 for Quiz B.

6. Quiz B should be covered up and Quiz C (two's and three's) completed.

7. Repeat steps 2 and 3 for Quiz C.

PAGE 140: CHAPTER QUIZ PAGE

It may be beneficial to complete this page on a separate sheet of paper to allow the learner to retake the quiz multiple times.

1. The learner completes this page with no guidance.

2. The instructor grades the quiz and it is recorded on the Score Sheet (p. 242).

3. If score is 80% or above, the certificate page (p. 243) should be completed; however, a lower score may indicate a need for more practice. If more practice is needed, the learner should redo chapter pages as well as spend more time practicing the song and completing game activities.

CHAPTER 5:
LEARNING THE FOUR'S

PAGES 142-144: INSTRUCTIONAL PAGES

Page 142: "Learning the Song"

1. Read the words of the "Four's Song" aloud without singing the tune. Have the learner read the words or repeat after you. Tell the learner to notice the picture clue. Explain that the tune of this song is "Oh My Darling Clementine."

2. Sing the song to the learner. (See page 17 for assistance with the song.) Explain that the picture shows a man singing about his girlfriend, Clementine. Count the strings on his guitar (four). Because there are four, the learner will remember that the "Four's Song" is to the tune of "Oh My Darling Clementine." Have the learner trace the four guitar strings in red.

3. Have the learner practice singing the song multiple times until it becomes very familiar. The learner should practice the song as much as possible. Consider singing while at recess or quietly during transition times. Also encourage the learner to practice at home, in the car or on the bus, while taking a walk, etc. Continue to remind the learner of the picture clue that accompanies the song.

Page 143: "Numbering the Fingers"

1. Tell the learner to place hands over the handprints on the page. Tell the learner to recall how the fingers are numbered. Review Chapter 2 if needed.

2. With hands on the handprints, have the learner sing the "Four's Song" aloud, remembering the picture clue. The learner should wiggle or touch down each finger in order while singing the song. Be sure to sing each word (number) of the song deliberately, pausing on the correct finger.

3. The learner will write the song in the blanks above the fingers. Remind the student of "counting by" a number as in Chapter 1, and that the song is the same as "counting by" fours.

4. The learner will complete the coloring portion of the page.

5. Have the learner practice singing the song while touching down the tips of the fingers to the paper. After the learner seems to understand the process, remove the worksheet and have him/her practice the same way, touching down fingers to any surface, such as a table, book, desk, or lap (even the air will work).

Page 144: "Learning to Multiply"

1. Have the learner look at Example A. The problem is 4 x 8 = __________.

2. Explain the following method to the learner: The first number in the problem is 4, so the "Four's Song" will be sung to find the answer. The second number in the problem is 8, so

the eighth finger is where the learner will stop singing. (The learner should imagine a miniature stop sign over the eighth finger, or one can even be made out of a bottle top or scrap of construction paper.) It is very important that the "stop" finger be identified and touched down. The number the learner is singing when he/she gets to the "stop" finger is the answer. In this case, the number being sung when the eighth finger touches down is thirty-two. Therefore, the answer is 32.

3. Follow this procedure to guide the learner through all examples on the page.

PAGES 145-146: PRACTICE PAGES

The learner will complete these pages with assistance from the instructor as needed. The instructor should monitor the learner's work to ensure the learner is using the finger-numbering method with songs and finger movements. Remind the learner that once a certain fact has been memorized, there is no need to continue using the method for that fact.

PAGES 147-148: ACTIVITY PAGES

Page 147: "Count by Cards"

1. Glue the page to a small piece of poster board or other heavy paper.
2. Cut on the lines.

3. Write the number four on the back of each card as a reminder that these cards go with the "Four's Song."

4. Use cards to play the following games or make up your own. Store cards in a resealable plastic bag or envelope, or secure with a rubber band.

Games:

Turn cards face up on a flat surface and mix up. Have the learner see how quickly he/she can put them in order. A kitchen timer or stopwatch can be used. Another option includes placing cards in order and practicing touching down fingers and singing songs.

Page 148: "Multiplication Cards"

1. Follow steps 1 and 2 from "Count by Cards."

2. Write answers on back of cards and store in resealable plastic bag or envelope, or secure with a rubber band.

3. Use cards to practice using the finger-numbering method and songs to multiply. Then check answers by flipping over cards to the back side. This game can be played individually or with multiple players.

PAGE 149: PRACTICE QUIZ PAGE

1. The learner completes Quiz A with no guidance from the instructor.

2. The instructor grades the paper by marking incorrect answers.

3. The learner should rework missed problems with guidance from the instructor.

4. Quiz A should be covered up with a folded sheet of paper or a book. Have the learner complete Quiz B. Be sure to point out that Quiz B contains numbers multiplied by three, thus a review of previously learned material.

5. Repeat steps 2 and 3 for Quiz B.

6. Quiz B should be covered up and Quiz C (three's and four's) completed.

7. Repeat steps 2 and 3 for Quiz C.

PAGE 150: CHAPTER QUIZ PAGE

It may be beneficial to complete this page on a separate sheet of paper to allow the learner to retake the quiz multiple times.

1. The learner completes this page with no guidance.

2. The instructor grades the quiz and it is recorded on the Score Sheet. (p. 242)

3. If score is 80% or above, the certificate page (p. 243) should be completed; however, a lower score may indicate a need for more practice. If more practice is needed, the learner should redo chapter pages as well as spend more time practicing the song and completing game activities.

CHAPTER 6:
LEARNING THE FIVE'S

PAGES 152-154: INSTRUCTIONAL PAGES

Page 152: "Learning the Song"

1. Read the words of the song aloud without singing the tune. Have the learner read the words or repeat after you. Tell the learner to notice the picture clue. Explain that the tune of this song is "Pop Goes the Weasel."

2. Sing the song to the learner. (See page 17 for assistance with the song.) Explain that the picture shows a weasel that is "popping" the number five. Because the five is being "popped," the learner will recall that the "Five's Song" is to the tune of "Pop Goes the Weasel." Have the learner color the number five red.

3. Have the learner practice singing the song multiple times until it becomes very familiar. The learner should practice the song as much as possible. Consider singing while at recess or quietly during transition times. Also encourage the learner to practice at home, in the car or on the bus, while taking a walk, etc. Continue to remind the learner of the picture clue that accompanies the song.

Page 153: "Numbering the Fingers"

1. Tell the learner to place hands over the handprints on the page. Tell the learner to recall how the fingers are numbered. Review Chapter 2 if needed.

2. With hands on the handprints, have the learner sing the "Five's Song" aloud, remembering the picture. The learner should wiggle or touch down each finger in order while singing the song. Be sure to sing each word (number) of the song deliberately, pausing on the correct finger.

3. The learner will write the song in the blanks above the fingers. Remind the student of "counting by" a number as in Chapter 1, and that the song is the same as "counting by" fives.

4. The learner will complete the coloring portion of the page.

5. Have the learner practice singing the song while touching down the tips of the fingers to the paper. After the learner seems to understand the process, remove the worksheet and have him/her practice the same way, touching down fingers to any surface, such as a table, book, desk, or lap (even the air will work).

Page 154: "Learning to Multiply"

1. Have the learner look at Example A. The problem is 5 x 6 = __________.

2. Explain the following method to the learner: The first number in the problem is 5, so the "Five's Song" will be sung to

find the answer. The second number in the problem is 6, so the sixth finger is where the learner will stop singing. (The learner should imagine a miniature stop sign over the sixth finger, or one can even be made out of a bottle top or scrap of construction paper.) It is very important that the "stop" finger be identified and touched down. The number the learner is singing when he/she gets to the "stop" finger is the answer. In this case, the number being sung when the sixth finger touches down is thirty. Therefore, the answer is 30.

3. Follow this procedure to guide the learner through all examples on the page.

PAGES 155-156: PRACTICE PAGES

The learner will complete these pages with assistance from the instructor as needed. The instructor should monitor the learner's work to ensure the learner is using the finger-numbering method with songs and finger movements. Remind the learner that once a certain fact has been memorized, there is no need to continue using the method for that fact.

PAGES 157-158: ACTIVITY PAGES

Page 157: "Count by Cards"

1. Glue the page to a small piece of poster board or other heavy paper.

2. Cut on the lines.

3. Write the number five on the back of each card as a reminder that these cards go with the "Five's Song."

4. Use cards to play the following games or make up your own. Store cards in a resealable plastic bag or envelope, or secure with a rubber band.

Games:

Turn cards face up on a flat surface and mix up. Have the learner see how quickly he/she can put them in order. A kitchen timer or stopwatch can be used. Another option includes placing cards in order and practicing touching down fingers and singing songs.

Page 158: "Multiplication Cards"

1. Follow steps 1 and 2 from "Count by Cards."

2. Write answers on back of cards and store in resealable plastic bag or envelope, or secure with a rubber band.

3. Use cards to practice using the finger-numbering method and songs to multiply. Then check answers by flipping over cards to the back side. This game can be played individually or with multiple players.

PAGE 159: PRACTICE QUIZ PAGE

1. The learner completes Quiz A with no guidance from the instructor.

2. The instructor grades the paper by marking incorrect answers.

3. The learner should rework missed problems with guidance from the instructor.

4. Quiz A should be covered up with a folded sheet of paper or a book. Have the learner complete Quiz B. Be sure to point out that Quiz B contains numbers multiplied by three and four, thus a review of previously learned material.

5. Repeat steps 2 and 3 for Quiz B.

6. Quiz B should be covered up and Quiz C (two's through five's) completed.

7. Repeat steps 2 and 3 for Quiz C.

PAGE 160: CHAPTER QUIZ PAGE

It may be beneficial to complete this page on a separate sheet of paper to allow the learner to retake the quiz multiple times.

1. The learner completes this page with no guidance.

2. The instructor grades the quiz and it is recorded on the Score Sheet. (p. 242)

3. If score is 80% or above, the certificate page (p. 243) should be completed; however, a lower score may indicate a need for more practice. If more practice is needed, the learner should redo chapter pages as well as spend more time practicing the song and completing game activities.

Important Note:

At this point you should determine whether or not to teach the learner that the order of the numbers in multiplication problems does not matter. Use your own judgment concerning the learner's level of understanding in deciding if this should be taught now or at a later time. If you feel he/she is ready, explain the following:

- If the learner already knows (has memorized) a multiplication fact, for example 2 x 5 = 10, then there is no need to use the song/method when given the problem 5 x 2 = 10. Illustrate two groups (circles) with five objects in each group. Have the learner count the total number of objects. The total is 10. Illustrate five groups (circles) with two objects in each group. Have the learner count the total number of objects. The total is 10. Explain that the answers (totals) are the same, or equal to one another, due to the fact that the order of numbers does not matter when multiplying. Give numerous examples.

- If the student knows one song better than the other, he/she can use the song that is more familiar. For example, if given 3 x 5 = __________, and the student is more comfortable with the "Five's Song" versus the "Three's Song," he/she can sing the "Five's Song" stopping at the third finger instead. Explain that it really does not matter whether the "Three's Song" or "Five's Song" is sung as long as you sing a song for one of the numbers and stop on the correct numbered finger for the other. Give numerous examples.

CHAPTER 7:
LEARNING THE ZERO'S AND ONE'S

PAGES 162-163: INSTRUCTIONAL PAGES

Page 162: "The Zero's Rule"

1. Explain to the learner that zeros are special, so they have a special rule. There will be no song or need for numbering the fingers.
2. Have the learner look at the boxes on the page for examples A-C. Ask how many apples are in each box. (There are none, or zero.) Tell the learner to write the number of apples in the blanks below each box (0, 0, 0...)
3. Ask the learner to add up the number of apples. (There are zero.) Tell the learner to write the answer in the blank to the right of the boxes in Examples A-C.
4. Have the learner to recall "counting by" numbers. He/she may use the songs learned thus far for the Two's through Five's. Review Chapter 1 if needed.
5. Ask the learner to now "count by zeros" (zero, zero, zero, zero.) Have the learner write the "count by's" in the blanks on Example D.

6. Explain that because of what's seen in the examples (adding groups of zero always equals zero, as well as counting by zero only produces answers of zeros) that any number multiplied by zero will equal zero.

7. Have the learner look at Example E-H. Point out the examples:

   0 x 4 = 0 3 x 0 = 0 0 x 9 = 0 6 x 0 = 0

8. Explain that anytime there is a zero in a multiplication problem, regardless of its position (coming first or second,) the answer will be zero. This is the special Zero's Rule.

Page 163: "The One's Magic"

1. Explain to the learner that like the zeros, the ones are also special and have their own rule as well. There will be no song; however, fingers can be used if needed.

2. Have the learner review "Counting by Ones" in Chapter 1. Look over and discuss those pages.

3. Tell the learner that no song is needed for the Ones because he/she already knows how to count by ones (one, two, three, four...). Have the learner demonstrate this by touching down the fingers.

4. Tell the learner to count by ones and write the answer in the blanks on Example A.

5. Have the learner look at Example B. The problem is 1 x 5 = 5. Explain that because the first number of the problem is 1, ordinarily the "One's Song" would be used, but there

is no such song. To have a song would be silly because the learner already knows how to count by one with no song (unlike the other numbers, where counting by a number isn't so easy.) Therefore, the learner thinks of counting by ones, with the stop sign over the fifth finger, and counts up. The number being said at the stop point is five.

6. Follow the above procedure for Example C. Have the learner write the answer.

7. Tell the learner that though he/she could count up each time, it would be much easier to use The One's Magic. Explain that the one in a problem is like magic (it even looks like a magic wand.) Whenever the learner sees the one (magic wand or magic one) in a problem, he/she can know that the answer will be whatever the other number is in the problem.

8. Have the learner look at Example D. Tell him/her to point to the magic wand (the one). The problem says 1 x 6 = _________. Then ask him/her to point to the other number in the problem. Explain that the magic wand (the one) magically puts a "spell" on the six to make it reappear as the answer. Have the learner write the answer in the blank.

9. Explain to the learner that where the one appears in the problem does not matter. It has magic powers regardless. Go over Examples E-H:

1 x 9 = 9 　　 7 x 1 = 7 　　 1 x 3 = 3 　　 5 x 1 = 5

PAGES 164-165: PRACTICE PAGES

The learner will complete these pages with assistance from the instructor as needed. The instructor should monitor the learner's work to ensure the learner is using the Zero's Rule or the One's Magic as needed.

Activity Pages

There are no activity pages for this chapter.

PAGE 166: PRACTICE QUIZ PAGE

1. The learner completes Quiz A with no guidance from the instructor (zero's only.)
2. The instructor grades the paper by marking incorrect answers.
3. The learner should rework missed problems with guidance from the instructor.
4. Quiz A should be covered up with a folded sheet of paper or a book. Have the learner complete Quiz B (one's only).
5. Repeat steps 2 and 3 for Quiz B.
6. Quiz B should be covered up and Quiz C completed (zero's and one's).
7. Repeat steps 2 and 3 for Quiz C.

PAGE 167: CHAPTER QUIZ PAGE

It may be beneficial to complete this page on a separate sheet of paper to allow the learner to retake the quiz multiple times.

1. The learner completes this page with no guidance.

2. The instructor grades the quiz and it is recorded on the Score Sheet (p. 242).

3. If score is 80% or above, the certificate page (p. 243) should be completed; however, a lower score may indicate a need for more practice. If more practice is needed, the learner should redo chapter pages as well as receive additional instruction.

CHAPTER 8:
LEARNING THE SIX'S

PAGES 170-172: INSTRUCTIONAL PAGES

Page 170: "Learning the Song"

1. Read the words of the song aloud without singing the tune. Have the learner read the words or repeat after you. Tell the learner to notice the picture clue. Explain that the tune of this song is "Camptown Races."

2. Sing the song to the learner. (See page 17 for assistance with the song.) Explain that the picture shows a girl running a race. The number on her shirt is six. Because there is a six on her shirt, the learner will be reminded that the "Six's Song" is to the tune of "Camptown Races." Have the learner color the square with the six red.

3. Have the learner practice singing the song multiple times until it becomes very familiar. The learner should practice the song as much as possible. Consider singing while at recess or quietly during transition times. Also encourage the learner to practice at home, in the car or on the bus, while taking a walk, etc. Continue to remind the learner of the picture clue that accompanies the song.

Page 171: "Numbering the Fingers"

1. Tell the learner to place hands over the handprints on the page. Tell the learner to recall how the fingers are numbered. Review Chapter 2 if needed.

2. With hands on the handprints, have the learner sing the "Six's Song" aloud, remembering the picture. The learner should wiggle or touch down each finger in order while singing the song. Be sure to sing each word (number) of the song deliberately, pausing on the correct finger.

3. The learner will write the song in the blanks above the fingers. Remind the student of "counting by" a number as in Chapter 1, and that the song is the same as "counting by" six.

4. The learner will complete the coloring portion of the page.

5. Have the learner practice singing the song while touching down the tips of the fingers to the paper. After the learner seems to understand the process, remove the worksheet and have him/her practice the same way, touching down fingers to any surface, such as a table, book, desk, or lap (even the air will work).

Page 172: "Learning to Multiply"

1. Have the learner look at Example A. The problem is 6 x 9 = __________.

2. Explain the following method to the learner: The first number in the problem is 6, so the "Six's Song" will be sung to

find the answer. The second number in the problem is 9, so the ninth finger is where the learner will stop singing. (The learner should imagine a miniature stop sign over the ninth finger, or one can even be made out of a bottle top or scrap of construction paper.) It is very important that the "stop" finger be identified and touched down. The number the learner is singing when he/she gets to the "stop" finger is the answer. In this case, the number being sung when the ninth finger touches down is fifty-four. Therefore, the answer is 54.

3. Follow this procedure to guide the learner through all examples on the page.

PAGES 173-174: PRACTICE PAGES

The learner will complete these pages with assistance from the instructor as needed. The instructor should monitor the learner's work to ensure the learner is using the finger-numbering method with songs and finger movements. Remind the learner that once a certain fact has been memorized, there is no need to continue using the method for that fact.

PAGES 175-176: ACTIVITY PAGES

Page 175: "Count by Cards"

1. Glue the page to a small piece of poster board or other heavy paper.
2. Cut on the lines.

3. Write the number six on the back of each card as a reminder that these cards go with the "Six's Song."
4. Use cards to play the following games or make up your own. Store cards in a resealable plastic bag or envelope, or secure with a rubber band.

Games:

Turn cards face up on a flat surface and mix up. Have the learner see how quickly he/she can put them in order. A kitchen timer or stopwatch can be used. Another option includes placing cards in order and practicing touching down fingers and singing songs.

Page 176 "Multiplication Cards"

1. Follow steps 1 and 2 from "Count by Cards."
2. Write answers on back of cards and store in resealable plastic bag or envelope, or secure with a rubber band.
3. Use cards to practice using the finger-numbering method and songs to multiply. Then check answers by flipping over cards to the back side. This game can be played individually or with multiple players.

PAGE 177: PRACTICE QUIZ PAGE

1. The learner completes Quiz A with no guidance from the instructor.

2. The instructor grades the paper by marking incorrect answers.

3. The learner should rework missed problems with guidance from the instructor.

4. Quiz A should be covered up with a folded sheet of paper or a book. Have the learner complete Quiz B. Be sure to point out that Quiz B contains numbers multiplied by both four and five, thus a review of previously learned material.

5. Repeat steps 2 and 3 for Quiz B.

6. Quiz B should be covered up and Quiz C (two's through six's) completed.

7. Repeat steps 2 and 3 for Quiz C.

PAGE 178: CHAPTER QUIZ PAGE

It may be beneficial to complete this page on a separate sheet of paper to allow the learner to retake the quiz multiple times.

1. The learner completes this page with no guidance.

2. The instructor grades the quiz and it is recorded on the Score Sheet. (p. 242)

3. If score is 80% or above, the certificate page (p. 243) should be completed; however, a lower score may indicate a need for more practice. If more practice is needed, the learner should redo chapter pages as well as spend more time practicing the song and completing game activities.

CHAPTER 9:
LEARNING THE SEVEN'S

PAGES 180-182: INSTRUCTIONAL PAGES

Page 180: "Learning the Song"

1. Read the words of the song aloud without singing the tune. Have the learner read the words or repeat after you. Tell the learner to notice the picture clue. Explain that the tune of this song is "Twinkle Twinkle Little Star."

2. Sing the song to the learner. (See page 17 for assistance with the song.) Explain that the picture shows stars in the sky in the form of a seven. Because they are in the form of the seven, the learner will recall that the "Seven's Song" is to the tune of "Twinkle Twinkle Little Star." Have the learner color the seven stars red.

3. Have the learner practice singing the song multiple times until it becomes very familiar. The learner should practice the song as much as possible. Consider singing while at recess or quietly during transition times. Also encourage the learner to practice at home, in the car or on the bus, while taking a walk, etc. Continue to remind the learner of the picture clue that accompanies the song.

Page 181: "Numbering the Fingers"

1. Tell the learner to place hands over the handprints on the page. Tell the learner to recall how the fingers are numbered. Review Chapter 2 if needed.

2. With hands on the handprints, have the learner sing the "Seven's Song" aloud, remembering the picture. The learner should wiggle or touch down each finger in order while singing the song. Be sure to sing each word (number) of the song deliberately, pausing on the correct finger.

3. The learner will write the song in the blanks above the fingers. Remind the student of "counting by" a number, as in Chapter 1, and that the song is the same as "counting by" seven.

4. The learner will complete the coloring portion of the page.

5. Have the learner practice singing the song while touching down the tips of the fingers to the paper. After the learner seems to understand the process, remove the worksheet and have him/her practice the same way, touching down fingers to any surface, such as a table, book, desk, or lap (even the air will work).

Page 182: "Learning to Multiply"

1. Have the learner look at Example A. The problem is 7 x 4 = ________.

2. Explain the following method to the learner: The first number in the problem is 7, so the "Seven's Song" will be sung to

find the answer. The second number in the problem is 4, so the fourth finger is where the learner will stop singing. (The learner should imagine a miniature stop sign over the fourth finger, or one can even be made out of a bottle top or scrap of construction paper.) It is very important that the "stop" finger be identified and touched down. The number the learner is singing when he/she gets to the "stop" finger is the answer. In this case, the number being sung when the fourth finger touches down is twenty-eight. Therefore, the answer is 28.

3. Follow this procedure to guide the learner through all examples on the page.

PAGES 183-184: PRACTICE PAGES

The learner will complete these pages with assistance from the instructor as needed. The instructor should monitor the learner's work to ensure the learner is using the finger-numbering method with songs and finger movements. Remind the learner that once a certain fact has been memorized, there is no need to continue using the method for that fact.

PAGES 185-186: ACTIVITY PAGES

Page 185: "Count by Cards"

1. Glue the page to a small piece of poster board or other heavy paper.
2. Cut on the lines.

3. Write the number seven on the back of each card as a reminder that these cards go with the "Seven's Song."

4. Use cards to play the following games or make up your own. Store cards in a resealable plastic bag or envelope, or secure with a rubber band.

Games:

Turn cards face up on a flat surface and mix up. Have the learner see how quickly he/she can put them in order. A kitchen timer or stopwatch can be used. Another option includes placing cards in order and practicing touching down fingers and singing songs.

Page 186: "Multiplication Cards"

1. Follow steps 1 and 2 from "Count by Cards."

2. Write answers on back of cards and store in resealable plastic bag or envelope, or secure with a rubber band.

3. Use cards to practice using the finger-numbering method and songs to multiply. Then check answers by flipping over cards to the back side. This game can be played individually or with multiple players.

PAGE 187: PRACTICE QUIZ PAGE

1. The learner completes Quiz A with no guidance from the instructor.

2. The instructor grades the paper by marking incorrect answers.

3. The learner should rework missed problems with guidance from the instructor.

4. Quiz A should be covered up with a folded sheet of paper or a book. Have the learner complete Quiz B. Be sure to point out that Quiz B contains numbers multiplied by five and six, thus a review of previously learned material.

5. Repeat steps 2 and 3 for Quiz B.

6. Quiz B should be covered up and Quiz C (zero's through seven's) completed.

7. Repeat steps 2 and 3 for Quiz C.

PAGE 188: CHAPTER QUIZ PAGE

It may be beneficial to complete this page on a separate sheet of paper to allow the learner to retake the quiz multiple times.

1. The learner completes this page with no guidance.

2. The instructor grades the quiz and it is recorded on the Score Sheet. (p. 242)

3. If score is 80% or above, the certificate page (p. 243) should be completed; however, a lower score may indicate a need for more practice. If more practice is needed, the learner should redo chapter pages as well as spend more time practicing the song and completing game activities.

CHAPTER 10:
LEARNING THE EIGHT'S

PAGES 190-192: INSTRUCTIONAL PAGES

Page 190: "Learning the Song"

1. Read the words of the song aloud without singing the tune. Have the learner read the words or repeat after you. Tell the learner to notice the picture clue. Explain that the tune of this song is "Mary Had A Little Lamb."

2. Sing the song to the learner. (See page 18 for assistance with the song.) Explain that the picture shows a girl, Mary, with her lamb. The lamb has had a good day, so Mary has given him eight treats. Because of the eight treats, the learner will remember that the "Eight's Song" is to the tune of "Mary Had A Little Lamb." Have the learner color the eight treats red.

3. Have the learner practice singing the song multiple times until it becomes very familiar. The learner should practice the song as much as possible. Consider singing while at recess or quietly during transition times. Also encourage the learner to practice at home, in the car or on the bus, while taking a

walk, etc. Continue to remind the learner of the picture clue that accompanies the song.

Page 191: "Numbering the Fingers"

1. Tell the learner to place hands over the handprints on the page. Tell the learner to recall how the fingers are numbered. Review Chapter 2 if needed.

2. With hands on the handprints, have the learner sing the "Eight's Song" aloud, remembering the picture. The learner should wiggle or touch down each finger in order while singing the song. Be sure to sing each word (number) of the song deliberately, pausing on the correct finger.

3. The learner will write the song in the blanks above the fingers. Remind the student of "counting by" a number as in Chapter 1 and that the song is the same as "counting by" eight.

4. The learner will complete the coloring portion of the page.

5. Have the learner practice singing the song while touching down the tips of the fingers to the paper. After the learner seems to understand the process, remove the worksheet and have him/her practice the same way, touching down fingers to any surface, such as a table, book, desk, or lap (even the air will work).

Page 192: "Learning to Multiply"

1. Have the learner look at Example A. The problem is 8 x 7 = ________.

2. Explain the following method to the learner:

The first number in the problem is 8, so the "Eight's Song" will be sung to find the answer. The second number in the problem is 7, so the seventh finger is where the learner will stop singing. (The learner should imagine a miniature stop sign over the seventh finger, or one can even be made out of a bottle top or scrap of construction paper). It is very important that the "stop" finger be identified and touched down. The number the learner is singing when he/she gets to the "stop" number is the answer. In this case, the number being sung when the seventh finger touches down is fifty-six. Therefore, the answer is 56.

3. Follow this procedure to guide the learner through all examples on the page.

PAGES 193-194: PRACTICE PAGES

The learner will complete these pages with assistance from the instructor as needed. The instructor should monitor the learner's work to ensure the learner is using the finger-numbering method with songs and finger movements. Remind the learner that once a certain fact has been memorized, there is no need to continue using the method for that fact.

PAGES 195-196: ACTIVITY PAGES

Page 195: "Count by Cards"

1. Glue the page to a small piece of poster board or other heavy paper.
2. Cut on the lines.
3. Write the number eight on the back of each card as a reminder that these cards go with the "Eight's Song."
4. Use cards to play the following games or make up your own. Store cards in a resealable plastic bag or envelope, or secure with a rubber band.

Games:

Turn cards face up on a flat surface and mix up. Have the learner see how quickly he/she can put them in order. A kitchen timer or stopwatch can be used. Another option includes placing cards in order and practice touching down fingers and singing songs.

Page 196: "Multiplication Cards"

1. Follow steps 1 and 2 from "Count by Cards."
2. Write answers on back of cards and store in resealable plastic bag or envelope, or secure with a rubber band.
3. Use cards to practice using the finger-numbering method and songs to multiply. Then check answers by flipping over cards to the back side. This game can be played individually or with multiple players.

PAGE 197: PRACTICE QUIZ PAGE

1. The learner completes Quiz A with no guidance from the instructor.
2. The instructor grades the paper by marking incorrect answers.
3. The learner should rework missed problems with guidance from the instructor.
4. Quiz A should be covered up with a folded sheet of paper or a book. Have the learner complete Quiz B. Be sure to point out that Quiz B contains numbers multiplied by six and seven, thus a review of previously learned material.
5. Repeat steps 2 and 3 for Quiz B.
6. Quiz B should be covered up and Quiz C (zero's through eight's) completed.
7. Repeat steps 2 and 3 for Quiz C.

PAGE 198: CHAPTER QUIZ PAGE

It may be beneficial to complete this page on a separate sheet of paper to allow the learner to retake the quiz multiple times.

1. The learner completes this page with no guidance.
2. The instructor grades the quiz and it is recorded on the Score Sheet. (p. 242)
3. If score is 80% or above, the certificate page (p. 243) should

be completed; however, a lower score may indicate a need for more practice. If more practice is needed, the learner should redo chapter pages as well as spend more time practicing the song and completing game activities.

CHAPTER 11:
LEARNING THE NINE'S

PAGES 200-201: INSTRUCTIONAL PAGES

Page 200: "Learning the Song"

1. Read the words of the song aloud without singing the tune. Have the learner read the words or repeat after you. Tell the learner to notice the picture clue. Explain that this song is to the tune of "For He's A Jolly Good Fellow."

2. Sing the song to the learner. (See page 18 for assistance with the song.) Explain that the picture shows a football player who has won the game, so his teammates are celebrating by singing "For He's A Jolly Good Fellow." The scoreboard shows that the team had nine points. Because of the nine on the scoreboard, the learner will remember that the "Nine's Song" is to the tune of "For He's A Jolly Good Fellow." With a red crayon, have the learner color the square containing the number nine.

3. Have the learner practice singing the song multiple times until it becomes very familiar. The learner should practice the song as much as possible. Consider singing while at recess or quietly during transition times. Also encourage the learner to practice at home, in the car or on the bus, while taking a

walk, etc. Continue to remind the learner of the picture clue that accompanies the song.

Page 201: "Numbering the Fingers"

1. Tell the learner to place hands over the handprints on the page. Tell the learner to recall how the fingers are numbered. Review Chapter 2 if needed.
2. With hands on the handprints, have the learner sing the "Nine's Song" aloud, remembering the picture. The learner should wiggle or touch down each finger in order while singing the song. Be sure to sing each word (number) of the song deliberately, pausing on the correct finger.
3. The learner will write the song in the blanks above the fingers. Remind the student of "counting by" a number, as in Chapter 1, and that the song is the same as "counting by" nine.
4. The learner will complete the coloring portion of the page.
5. Have the learner practice singing the song while touching down the tips of the fingers to the paper. After the learner seems to understand the process, remove the worksheet and have him/her practice the same way, touching down fingers to any surface, such as a table, book, desk, or lap (even the air will work).

Page 202: "Learning to Multiply"

1. Have the learner look at Example A. The problem is 9 x 3 = _________.

2. Explain the following method to the learner: The first number in the problem is 9, so the "Nine's Song" will be sung to find the answer. The second number in the problem is 3, so the third finger is where the learner will stop singing. (The learner should imagine a miniature stop sign over the third finger, or one can even be made out of a bottle top or scrap of construction paper.) It is very important that the "stop" finger be identified and touched down. The number the learner is singing when he/she gets to the "stop" finger is the answer. In this case, the number being sung when the third finger touches down is twenty-seven. Therefore, the answer is 27.

3. Follow this procedure to guide the learner through all examples on the page.

PAGES 203-204: PRACTICE PAGES

The learner will complete these pages with assistance from the instructor as needed. The instructor should monitor the learner's work to ensure the learner is using the finger-numbering method with songs and finger movements. Remind the learner that once a certain fact has been memorized, there is no need to continue using the method for that fact.

PAGES 205-206: ACTIVITY PAGES

Page 205: "Count by Cards"

1. Glue the page to a small piece of poster board or other heavy paper.

2. Cut on the lines.
3. Write the number nine on the back of each card as a reminder that these cards go with the "Nine's Song."
4. Use cards to play the following games or make up your own. Store cards in a resealable plastic bag or envelope, or secure with a rubber band.

Games:

Turn cards face up on a flat surface and mix up. Have the learner see how quickly he/she can put them in order. A kitchen timer or stopwatch can be used. Another option includes placing cards in order and practicing touching down fingers and singing songs.

Page 206: "Multiplication Cards"

1. Follow steps 1 and 2 from "Count by Cards."
2. Write answers on back of cards and store in plastic bag or envelope, or secure with a rubber band.
3. Use cards to practice using the finger-numbering method and songs to multiply. Then check answers by flipping over cards to the back side. This game can be played individually or with multiple players.

PAGE 207: PRACTICE QUIZ PAGE

1. The learner completes Quiz A with no guidance from the instructor.

2. The instructor grades the paper by marking incorrect answers.

3. The learner should rework missed problems with guidance from the instructor.

4. Quiz A should be covered up with a folded sheet of paper or a book. Have the learner complete Quiz B. Be sure to point out that Quiz B contains numbers multiplied by seven and eight, thus a review of previously learned material.

5. Repeat steps 2 and 3 for Quiz B.

6. Quiz B should be covered up and Quiz C (zero's through nine's) completed.

7. Repeat steps 2 and 3 for Quiz C.

PAGE 208: CHAPTER QUIZ PAGE

It may be beneficial to complete this page on a separate sheet of paper to allow the learner to retake the quiz multiple times.

1. The learner completes this page with no guidance.

2. The instructor grades the quiz and it is recorded on the Score Sheet (p. 242).

3. If score is 80% or above, the certificate page (p. 243) should be completed; however, a lower score may indicate a need for more practice. If more practice is needed, the learner should redo chapter pages as well as spend more time practicing the song and completing game activities.

CHAPTER 12:
LEARNING THE TEN'S

PAGES 210-211: INSTRUCTIONAL PAGES

Page 210: "Learning the Song"

1. Read the words of the song aloud without singing the tune. Have the learner read the words or repeat after you. Tell the learner to notice the picture clue. Explain that this song is to the tune of "Three Blind Mice."

2. Sing the song to the learner. (See page 18 for assistance with the song.) Explain that the picture shows three mice about to enter a door. The door has a ten on it. Because of the ten on the door that the mice are about to go into, the learner will remember that the "Ten's Song" is to the tune of "Three Blind Mice." Have the learner color the square house number red.

3. Have the learner practice singing the song multiple times until it becomes very familiar. The learner should practice the song as much as possible. Consider singing while at recess or quietly during transition times. Also encourage the learner to practice at home, in the car or on the bus, while taking a

walk, etc. Continue to remind the learner of the picture that accompanies the song.

Page 211: "Numbering the Fingers"

1. Tell the learner to place hands over the handprints on the page. Tell the learner to recall how the fingers are numbered. Review Chapter 2 if needed.
2. With hands on the handprints, have the learner sing the "Ten's Song" aloud, remembering the picture clue. The learner should wiggle or touch down each finger in order while singing the song. Be sure to sing each word (number) of the song deliberately, pausing on the correct finger.
3. The learner will write the song in the blanks above the fingers. Remind the student of "counting by" a number as in Chapter 1, and that the song is the same as "counting by" tens.
4. The learner will complete the coloring portion of the page.
5. Have the learner practice singing the song while touching down the tips of the fingers to the paper. After the learner seems to understand the process, remove the worksheet and have him/her practice the same way, touching down fingers to any surface, such as a table, book, desk, or lap (even the air will work.)

Page 212: "Learning to Multiply"

1. Have the learner look at Example A. The problem is 10 x 6 = __________.

2. Explain the following method to the learner: The first number in the problem is 10, so the "Ten's Song" will be sung to find the answer. The second number in the problem is 6, so the sixth finger is where the learner will stop singing. (The learner should imagine a miniature stop sign over the sixth finger, or one can even be made out of a bottle top or scrap of construction paper.) It is very important that the "stop" finger be identified and touched down. The number the learner is singing when he/she gets to the "stop" finger is the answer. In this case, the number being sung when the sixth finger touches down is sixty. Therefore, the answer is 60.

3. Follow this procedure to guide the learner through all examples on the page.

PAGES 213-214: PRACTICE PAGES

The learner will complete these pages with assistance from the instructor as needed. The instructor should monitor the learner's work to ensure the learner is using the finger-numbering method with songs and finger movements. Remind the learner that once a certain fact has been memorized, there is no need to continue using the method for that fact.

PAGES 215-216: ACTIVITY PAGES

Page 215: "Count by Cards"

1. Glue the page to a small piece of poster board or other heavy paper.

2. Cut on the lines.
3. Write the number ten on the back of each card as a reminder that these cards go with the "Ten's Song."
4. Use cards to play the following games or make up your own. Store cards in a resealable plastic bag or envelope, or secure with a rubber band.

Games:

Turn cards face up on a flat surface and mix up. Have the learner see how quickly he/she can put them in order. A kitchen timer or stopwatch can be used. Another option includes placing cards in order and practicing touching down fingers and singing songs.

Page 216: "Multiplication Cards"

1. Follow steps 1 and 2 from "Count by Cards."
2. Write answers on back of cards and store in resealable plastic bag or envelope, or secure with a rubber band.
3. Use cards to practice using the finger-numbering method and songs to multiply. Then check answers by flipping over cards to the back side. This game can be played individually or with multiple players.

PAGE 217: PRACTICE QUIZ PAGE

1. The learner completes Quiz A with no guidance from the instructor.

2. The instructor grades the paper by marking incorrect answers.

3. The learner should rework missed problems with guidance from the instructor.

4. Quiz A should be covered up with a folded sheet of paper or a book. Have the learner complete Quiz B. Be sure to point out that Quiz B contains numbers multiplied by eight and nine, thus a review of preciously learned material.

5. Repeat steps 2 and 3 for Quiz B.

6. Quiz B should be covered up and Quiz C (zero's through ten's) completed.

7. Repeat steps 2 and 3 for Quiz C.

PAGE 218: CHAPTER QUIZ PAGE

It may be beneficial to complete this page on a separate sheet of paper to allow the learner to retake the quiz multiple times.

1. The learner completes this page with no guidance.

2. The instructor grades the quiz and it is recorded on the Score Sheet (p. 242).

3. If score is 80% or above, the certificate page (p. 243) should be completed; however, a lower score may indicate a need for more practice. If more practice is needed, the learner should redo chapter pages as well as spend more time practicing the song and completing game activities.

CHAPTER 13: LEARNING THE ELEVEN'S

PAGE 220: INSTRUCTIONAL PAGE

Page 220: "The Eleven's Trick"

1. Explain to the learner that like the zero's and one's, the eleven's are also special and have their own rule as well. There will be no song.

2. Have the learner review "counting by a number" as discussed in Chapter 1. Look over and review those pages.

3. Have the learner use repeated addition to count by elevens and write the answers in Example A. Use a separate sheet of paper if needed.

4. Tell the learner that though he/she could count up by eleven each time, it would be much easier to use "The Eleven's Trick." Remind the learner of the "One's Magic." (The one in a problem is like magic—it even looks like a magic wand. Whenever the learner saw the one (magic wand or magic one) in a problem, he/she knew that the answer would be whatever the other number was in the problem.) Just like the one was "magic," the eleven is also. Because an eleven is made up of two ones (wands), they can perform a trick.

5. Have the learner look at Example B. Tell him/her to point to the magic wands (the ones). The problem says 11 x 5 = 55. Then ask him/her to point to the other number in the problem (the five). Explain that the magic wands (the ones) magically "trick" the fives into reappearing twice as the answer (55).

6. Tell the learner to look at Example C. Have him/her point to the magic wands (eleven). Ask the learner to identify the other number in the problem (eight). Tell the learner to imagine the wands performing the "trick" on the eight. Have the learner write the answer in the blank.

7. Explain to the learner that where the eleven appears in the problem does not matter. It has powers to perform the "trick" regardless.

8. Go over Examples D-H.

   11 x 6 = 66   11 x 9 = 99   7 x 11 = 77   11 x 3 =33   5 x 11 = 55

9. Explain that 11 x 10 is an exception to the "trick" and that it makes itself reappear along with a zero (110.)

PAGES 221-222: PRACTICE PAGES

The learner will complete these pages with assistance from the instructor as needed. The instructor should monitor the learner's work to ensure the learner is using the "Eleven's Trick" as needed.

ACTIVITY PAGES

There are no activity pages for this chapter.

PAGE 223: PRACTICE QUIZ PAGE

1. The learner completes Quiz A with no guidance from the instructor (eleven's only.)
2. The instructor grades the paper by marking incorrect answers.
3. The learner should rework missed problems with guidance from the instructor.
4. Quiz A should be covered up with a folded sheet of paper or a book. Have the learner complete Quiz B. Be sure to point out that Quiz B contains numbers multiplied by nine and ten, thus a review of previously learned material.
5. Repeat steps 2 and 3 for Quiz B.
6. Quiz B should be covered up and Quiz C completed (zero's through eleven's).
7. Repeat steps 2 and 3 for Quiz C.

PAGE 224: CHAPTER QUIZ PAGE

It may be beneficial to complete this page on a separate sheet of paper to allow the learner to retake the quiz multiple times.

1. The learner completes this page with no guidance.
2. The instructor grades the quiz and it is recorded on the Score Sheet (p. 242).
3. If score is 80% or above, the certificate page (p. 243) should

be completed; however, a lower score may indicate a need for more practice. If more practice is needed, the learner should redo chapter pages as well as receive additional instruction.

CHAPTER 14:
LEARNING THE TWELVE'S

PAGES 226-228: INSTRUCTIONAL PAGES

Page 226: "Learning the Song"

1. Read the words of the song aloud without singing the tune. Have the learner read the words or repeat after you. Tell the learner to notice the picture clue. Explain that this song is to the tune of "Billy Boy."

2. Sing the song to the learner. (See page 18 for assistance with the song.) Explain that the picture shows a mother talking to her son, Billy about "where he's been." The address of the house is number twelve. Because of the twelve on the mailbox, the learner will remember that the "Twelve's Song" is to the tune of "Billy Boy." Have the learner color the mailbox red.

3. Have the learner practice singing the song multiple times until it becomes very familiar. The learner should practice the song as much as possible. Consider singing while at recess or quietly during transition times. Also encourage the learner to practice at home, in the car or on the bus, while taking a walk, etc. Continue to remind the learner of the picture that accompanies the song.

Page 227: "Numbering the Fingers"

1. Tell the learner to place hands over the handprints on the page. Tell the learner to recall how the fingers are numbered. Review Chapter 2 if needed.

2. With hands on the handprints, have the learner sing the "Twelve's Song" aloud, remembering the picture clue. The learner should wiggle or touch down each finger in order while singing the song. Be sure to sing each word (number) of the song deliberately, pausing on the correct finger.

3. The learner will write the song in the blanks above the fingers. Remind the student of "counting by" a number as in Chapter 1, and that the song is the same as "counting by."

4. The learner will complete the coloring portion of the page.

5. Have the learner practice singing the song while touching down the tips of the fingers to the paper. After the learner seems to understand the process, remove the workbook and have him/her practice the same way, touching down fingers to any surface, such as a table, book, desk, or lap (even the air will work).

Page 228: "Learning to Multiply"

1. Have the learner look at Example A. The problem is 12 x 6 __________.

2. Explain the following method to the learner: The first number in the problem is 12, so the "Twelve's Song" will be sung

to find the answer. The second number in the problem is 6, so the sixth finger is where the learner will stop singing. (The learner should imagine a miniature stop sign over the sixth finger, or one can even be made out of a bottle top or scrap of construction paper.) It is very important that the "stop" finger be identified and touched down. The number the learner is singing when he/she gets to the "stop" finger is the answer. In this case, the number being sung when the sixth finger touches down is seventy-two. Therefore, the answer is 72.

3. Follow this procedure to guide the learner through all examples on the page.

PAGES 229-230: PRACTICE PAGES

The learner will complete these pages with assistance from the instructor as needed. The instructor should monitor the learner's work to ensure the learner is using the finger-numbering method with songs and finger movements. Remind the learner that once a certain fact has been memorized, there is no need to continue using the method for that fact.

PAGES 231-232: ACTIVITY PAGES

Page 231: "Count by Cards"

1. Glue the page to a small piece of poster board or other heavy paper.
2. Cut on the lines.

3. Write the number twelve on the back of each card as a reminder that these cards go with the "Twelve's Song."

4. Use cards to play the following games or make up your own. Store cards in a resealable plastic bag or envelope, or secure with a rubber band.

Games:

Turn cards face up on a flat surface and mix up. Have the learner see how quickly he/she can put them in order. A kitchen timer or stopwatch can be used. Another option includes placing cards in order and practicing touching down fingers and singing songs.

Page 232: "Multiplication Cards"

1. Follow steps 1 and 2 from "Count by Cards."

2. Write answers on back of cards and store in resealable plastic bag or envelope, or secure with a rubber band.

3. Use cards to practice using the finger-numbering method and songs to multiply. Then check answers by flipping over cards to the back side. This game can be played individually or with multiple players.

PAGE 233: PRACTICE QUIZ PAGE

1. The learner completes Quiz A with no guidance from the instructor.

2. The instructor grades the paper by marking incorrect answers.

3. The learner should rework missed problems with guidance from the instructor.

4. Quiz A should be covered up with a folded sheet of paper or a book. Have the learner complete Quiz B. Be sure to point out that Quiz B contains numbers multiplied by ten and eleven, thus a review of previously learned material.

5. Repeat steps 2 and 3 for Quiz B.

6. Quiz B should be covered up and Quiz C (zero's through twelve's) completed.

7. Repeat steps 2 and 3 for Quiz C.

PAGE 234: CHAPTER QUIZ PAGE

It may be beneficial to complete this page on a separate sheet of paper to allow the learner to retake the quiz multiple times.

1. The learner completes this page with no guidance.

2. The instructor grades the quiz and it is recorded on the Score Sheet (p. 242).

3. If score is 80% or above, the certificate page (p. 243) should be completed; however, a lower score may indicate a need for more practice. If more practice is needed, the learner should redo chapter pages as well as spend more time practicing the song and completing game activities.

CHAPTER 15:
EXTRA PRACTICE

PAGES 236-239: PRACTICE PAGES

This chapter includes pages of additional practice. They can be used for homework, classwork, quizzes, tests, etc. These pages can be presented a number of times, well after initial instruction to ensure that the learners continue to remember and practice the method. You should be able to find practice pages such as these from a variety of other resources as well.

STUDENT WORKSHEETS

STUDENT WORKSHEETS: CHAPTER 1

COUNTING BY A NUMBER

Chapter 1: Instructional Page Counting by One

Count by one.

1	2	3	4	5
6	7	8	9	10

Count by one.

___	___	___	___	___
___	___	___	___	___

Count by one.

___ ___ ___ ___ ___

___ ___ ___ ___ ___

Count by one.

Chapter 1: Instructional Page — Counting by Two

Count by two.

❑❑ 2	❑❑ 4	❑❑ 6	❑❑ 8	❑❑ 10
❑❑ 12	❑❑ 14	❑❑ 16	❑❑ 18	❑❑ 20

Count by two.

☼☼ ____	☼☼ ____	☼☼ ____	☼☼ ____	☼☼ ____
☼☼ ____	☼☼ ____	☼☼ ____	☼☼ ____	☼☼ ____

Count by two.

_____ _____ _____ _____ _____

_____ _____ _____ _____ _____

Count by two.

Count by three.

☺☺☺ 3	☺☺☺ 6	☺☺☺ 9	☺☺☺ 12	☺☺☺ 15
☺☺☺ 18	☺☺☺ 21	☺☺☺ 24	☺☺☺ 27	☺☺☺ 30

Count by three.

♦♦♦ ____	♦♦♦ ____	♦♦♦ ____	♦♦♦ ____	♦♦♦ ____
♦♦♦ ____	♦♦♦ ____	♦♦♦ ____	♦♦♦ ____	♦♦♦ ____

Count by three.

____ ____ ____ ____ ____

____ ____ ____ ____ ____

Count by three.

Chapter 1: Instructional Page Counting by Four

Count by four.

🏳🏳🏳🏳 4	🏳🏳🏳🏳 8	🏳🏳🏳🏳 12	🏳🏳🏳🏳 16	🏳🏳🏳🏳 20
🏳🏳🏳🏳 24	🏳🏳🏳🏳 28	🏳🏳🏳🏳 32	🏳🏳🏳🏳 36	🏳🏳🏳🏳 40

Count by four.

☼☼☼☼ ______	☼☼☼☼ ______	☼☼☼☼ ______	☼☼☼☼ ______	☼☼☼☼ ______
☼☼☼☼ ______	☼☼☼☼ ______	☼☼☼☼ ______	☼☼☼☼ ______	☼☼☼☼ ______

Count by four.

______ ______ ______ ______ ______

______ ______ ______ ______ ______

Count by four.

Count by five.

☺☺☺☺☺ 5	☺☺☺☺☺ 10	☺☺☺☺☺ 15	☺☺☺☺☺ 20	☺☺☺☺☺ 25
☺☺☺☺☺ 30	☺☺☺☺☺ 35	☺☺☺☺☺ 40	☺☺☺☺☺ 45	☺☺☺☺☺ 50

Count by five.

❑❑❑❑❑ ____	❑❑❑❑❑ ____	❑❑❑❑❑ ____	❑❑❑❑❑ ____	❑❑❑❑❑ ____
❑❑❑❑❑ ____	❑❑❑❑❑ ____	❑❑❑❑❑ ____	❑❑❑❑❑ ____	❑❑❑❑❑ ____

Count by five.

____ ____ ____ ____ ____

____ ____ ____ ____ ____

Count by five.

Chapter 1: Instructional Page — Counting by Six

Count by six.

6	12	18	24	30
36	42	48	54	60

Count by six.

____	____	____	____	____
____	____	____	____	____

Count by six.

____ ____ ____ ____ ____

____ ____ ____ ____ ____

Count by six.

Chapter 1: Instructional Page Counting by Seven

Count by seven.

❑❑❑❑❑❑❑	❑❑❑❑❑❑❑	❑❑❑❑❑❑❑	❑❑❑❑❑❑❑	❑❑❑❑❑❑❑
7	14	21	28	35
❑❑❑❑❑❑❑	❑❑❑❑❑❑❑	❑❑❑❑❑❑❑	❑❑❑❑❑❑❑	❑❑❑❑❑❑❑
42	49	56	63	70

Count by seven.

❀❀❀❀❀❀❀	❀❀❀❀❀❀❀	❀❀❀❀❀❀❀	❀❀❀❀❀❀❀	❀❀❀❀❀❀❀
____	____	____	____	____
❀❀❀❀❀❀❀	❀❀❀❀❀❀❀	❀❀❀❀❀❀❀	❀❀❀❀❀❀❀	❀❀❀❀❀❀❀
____	____	____	____	____

Count by seven.

____ ____ ____ ____ ____

____ ____ ____ ____ ____

Count by seven.

Count by eight.

❑❑❑❑ ❑❑❑❑ 8	❑❑❑❑ ❑❑❑❑ 16	❑❑❑❑ ❑❑❑❑ 24	❑❑❑❑ ❑❑❑❑ 32	❑❑❑❑ ❑❑❑❑ 40
❑❑❑❑ ❑❑❑❑ 48	❑❑❑❑ ❑❑❑❑ 56	❑❑❑❑ ❑❑❑❑ 64	❑❑❑❑ ❑❑❑❑ 72	❑❑❑❑ ❑❑❑❑ 80

Count by eight.

♦♦♦♦ ♦♦♦♦ ____	♦♦♦♦ ♦♦♦♦ ____	♦♦♦♦ ♦♦♦♦ ____	♦♦♦♦ ♦♦♦♦ ____	♦♦♦♦ ♦♦♦♦ ____
♦♦♦♦ ♦♦♦♦ ____	♦♦♦♦ ♦♦♦♦ ____	♦♦♦♦ ♦♦♦♦ ____	♦♦♦♦ ♦♦♦♦ ____	♦♦♦♦ ♦♦♦♦ ____

Count by eight.

____ ____ ____ ____ ____

____ ____ ____ ____ ____

Count by eight.

Chapter 1: Instructional Page Counting by Nine

Count by nine.

9	18	27	36	45
54	63	72	81	90

Count by nine.

______	______	______	______	______
______	______	______	______	______

Count by nine.

______ ______ ______ ______ ______

______ ______ ______ ______ ______

Count by nine.

Count by ten.

☺☺☺☺☺ ☺☺☺☺☺ 10	☺☺☺☺☺ ☺☺☺☺☺ 20	☺☺☺☺☺ ☺☺☺☺☺ 30	☺☺☺☺☺ ☺☺☺☺☺ 40	☺☺☺☺☺ ☺☺☺☺☺ 50
☺☺☺☺☺ ☺☺☺☺☺ 60	☺☺☺☺☺ ☺☺☺☺☺ 70	☺☺☺☺☺ ☺☺☺☺☺ 80	☺☺☺☺☺ ☺☺☺☺☺ 90	☺☺☺☺☺ ☺☺☺☺☺ 100

Count by ten.

❑❑❑❑❑ ❑❑❑❑❑ ______	❑❑❑❑❑ ❑❑❑❑❑ ______	❑❑❑❑❑ ❑❑❑❑❑ ______	❑❑❑❑❑ ❑❑❑❑❑ ______	❑❑❑❑❑ ❑❑❑❑❑ ______
❑❑❑❑❑ ❑❑❑❑❑ ______	❑❑❑❑❑ ❑❑❑❑❑ ______	❑❑❑❑❑ ❑❑❑❑❑ ______	❑❑❑❑❑ ❑❑❑❑❑ ______	❑❑❑❑❑ ❑❑❑❑❑ ______

Count by ten.

______ ______ ______ ______ ______

______ ______ ______ ______ ______

Count by ten.

Count by eleven.

♦♦♦♦♦♦ ♦♦♦♦♦ 11	♦♦♦♦♦♦ ♦♦♦♦♦ 22	♦♦♦♦♦♦ ♦♦♦♦♦ 33	♦♦♦♦♦♦ ♦♦♦♦♦ 44	♦♦♦♦♦♦ ♦♦♦♦♦ 55
♦♦♦♦♦♦ ♦♦♦♦♦ 66	♦♦♦♦♦♦ ♦♦♦♦♦ 77	♦♦♦♦♦♦ ♦♦♦♦♦ 88	♦♦♦♦♦♦ ♦♦♦♦♦ 99	♦♦♦♦♦♦ ♦♦♦♦♦ 110

Count by eleven.

❑❑❑❑❑❑ ❑❑❑❑❑ ____	❑❑❑❑❑❑ ❑❑❑❑❑ ____	❑❑❑❑❑❑ ❑❑❑❑❑ ____	❑❑❑❑❑❑ ❑❑❑❑❑ ____	❑❑❑❑❑❑ ❑❑❑❑❑ ____
❑❑❑❑❑❑ ❑❑❑❑❑ ____	❑❑❑❑❑❑ ❑❑❑❑❑ ____	❑❑❑❑❑❑ ❑❑❑❑❑ ____	❑❑❑❑❑❑ ❑❑❑❑❑ ____	❑❑❑❑❑❑ ❑❑❑❑❑ ____

Count by eleven.

____ ____ ____ ____ ____

____ ____ ____ ____ ____

Count by eleven.

Chapter 1: Instructional Page Counting by Twelve

Count by twelve.

❑❑❑❑❑❑ ❑❑❑❑❑❑ 12	❑❑❑❑❑❑ ❑❑❑❑❑❑ 24	❑❑❑❑❑❑ ❑❑❑❑❑❑ 36	❑❑❑❑❑❑ ❑❑❑❑❑❑ 48	❑❑❑❑❑❑ ❑❑❑❑❑❑ 60
❑❑❑❑❑❑ ❑❑❑❑❑❑ 72	❑❑❑❑❑❑ ❑❑❑❑❑❑ 84	❑❑❑❑❑❑ ❑❑❑❑❑❑ 96	❑❑❑❑❑❑ ❑❑❑❑❑❑ 108	❑❑❑❑❑❑ ❑❑❑❑❑❑ 120

Count by twelve.

______	______	______	______	______
______	______	______	______	______

Count by twelve.

______ ______ ______ ______ ______

______ ______ ______ ______ ______

Count by twelve.

STUDENT WORKSHEETS: CHAPTER 2

NUMBERING THE FINGERS

Chapter 2: Instructional Page

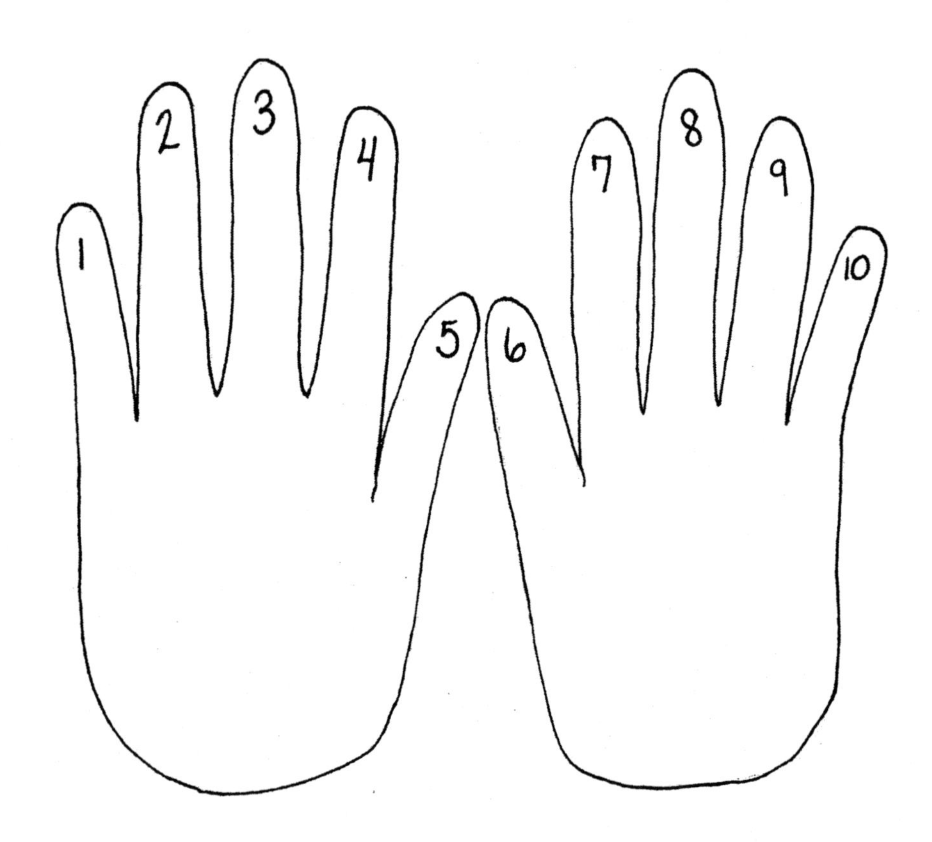

Color the fingers above.

1- red	2- blue	3- brown	4- yellow	5- purple
6- black	7- green	8- pink	9- orange	10- gray

Chapter 2: Practice Page

A. Draw a line from the finger to the box with the correct number.

8	1	6	5	2	10	3	7	4	9

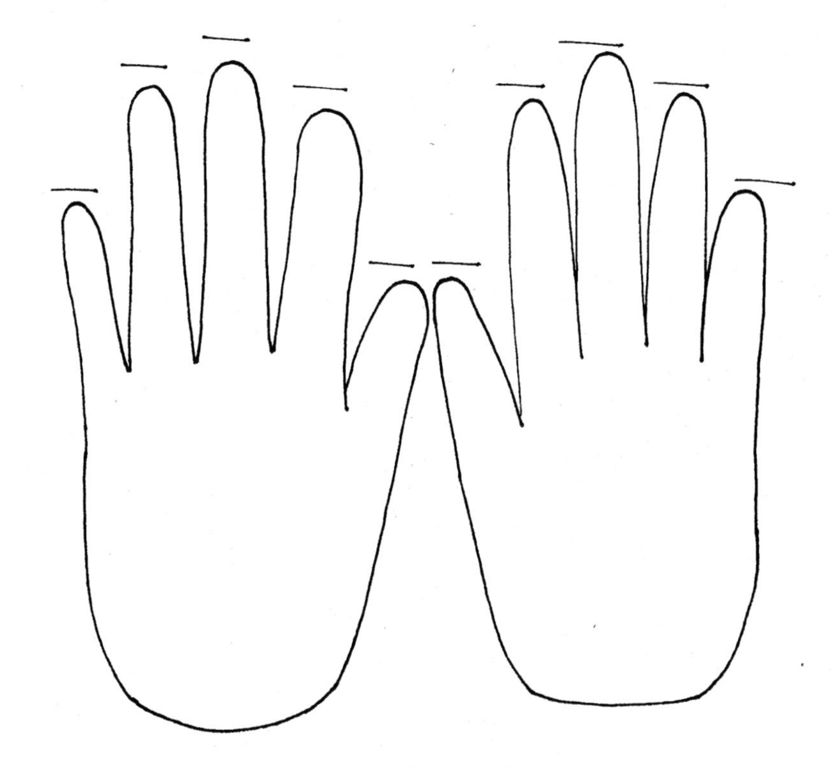

B. Color the fingers above.

1- yellow	2- green	3- pink	4- brown	5- gray
6- blue	7- red	8- orange	9- purple	10- black

C. Write the correct numbers in the blanks.

STUDENT WORKSHEETS: CHAPTER 3

LEARNING THE TWO'S

Chapter 3: Learning the Song

The "Two's Song"
(To the tune of "The Farmer in the Dell")

Two, Four, and Six
Eight, Ten, and Twelve
Fourteen, Sixteen,
Eighteen, and Twenty.

2, 4, and 6
8, 10, and 12
14, 16,
18, and 20.

Chapter 3: Numbering the Fingers

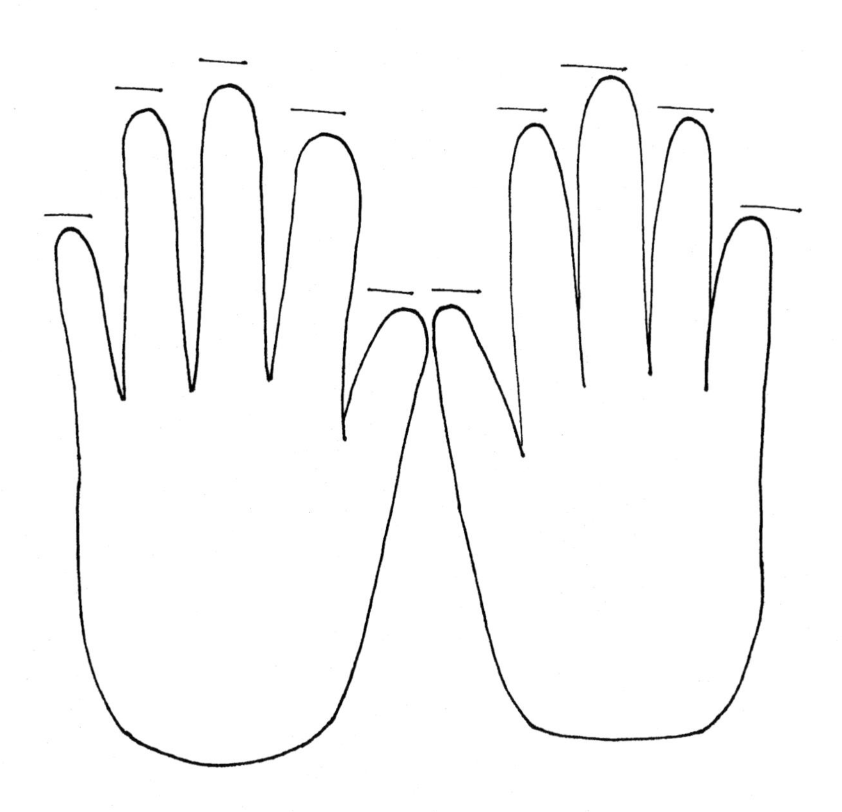

Color the fingers.

1- blue	4- yellow	6- red	3- gray	10- purple
2- green	9- orange	5- black	7- brown	8-pink

Chapter 3: Learning to Multiply

Example A: $2 \times 4 =$ ____

The first number is 2. (Sing the "Two's Song.")
The second number is 4. (Stop on the fourth finger.)
The answer is 8. (The word of the song you're singing when you stop.)

Example B: $2 \times 9 =$ ____

The first number is 2. (Sing the "Two's Song.")
The second number is 9. (Stop on the ninth finger.)
The answer is 18. (The word of the song you're singing when you stop.)

Example C: $2 \times 7 =$ ____

The first number is 2. (Sing the "Two's Song.")
The second number is 7. (Stop on the seventh finger.)
The answer is 14. (The word of the song you're singing when you stop.)

Chapter 3: Practice Page

Draw the picture clue that makes you think of the "Two's Song."
Write the "Two's Song" in the blanks. ____ ____ ____ ____ ____ ____ ____ ____ ____ ____

1. 2 x 4 = _____	1. 2 x 7 = _____
2. 2 x 9 = _____	2. 2 x 2 = _____
3. 2 x 7 = _____	3. 2 x 6 = _____
4. 2 x 8 = _____	4. 2 x 10 = _____
5. 2 x 5 = _____	5. 2 x 1 = _____
6. 2 x 6 = _____	6. 2 x 9 = _____
7. 2 x 1 = _____	7. 2 x 8 = _____
8. 2 x 10 = _____	8. 2 x 4 = _____
9. 2 x 3 = _____	9. 2 x 5 = _____
10. 2 x 2 = _____	10. 2 x 3 = _____

Chapter 3: Practice Page

1. 2 x 8 = _____	11. 2 x 1 = _____
2. 2 x 9 = _____	12. 2 x 5 = _____
3. 2 x 3 = _____	13. 2 x 4 = _____
4. 2 x 6 = _____	14. 2 x 8 = _____
5. 2 x 1 = _____	15. 2 x 6 = _____
6. 2 x 10 = _____	16. 2 x 9 = _____
7. 2 x 5 = _____	17. 2 x 7 = _____
8. 2 x 4 = _____	18. 2 x 3 = _____
9. 2 x 7 = _____	19. 2 x 2 = _____
10. 2 x 2 = _____	20. 2 x 10 = _____

Write the "Two's Song" in the blanks.

___ ___ ___ ___ ___ ___ ___ ___ ___ ___

Chapter 3: Count by Cards

2	4
6	8
10	12
14	16
18	20

Chapter 3: Multiplication Cards

2 x 1	2 x 2
2 x 3	2 x 4
2 x 5	2 x 6
2 x 7	2 x 8
2 x 9	2x10

Chapter 3: Practice Quiz

Quiz A

1. 2 x 2 = ____	6. 2 x 5 = ____
2. 2 x 4 = ____	7. 2 x 10= ____
3. 2 x 1 = ____	8. 2 x 7 = ____
4. 2 x 9 = ____	9. 2 x 3 = ____
5. 2 x 6 = ____	10. 2 x 8 = ____

Quiz B

1. 2 x 5 = ____	6. 2 x 7 = ____
2. 2 x 1 = ____	7. 2 x 2 = ____
3. 2 x 6 = ____	8. 2 x 3 = ____
4. 2 x 10= ____	9. 2 x 9 = ____
5. 2 x 8= ____	10. 2 x 4 = ____

Quiz C

1. 2 x 4 = ____	6. 2 x 8 = ____
2. 2 x 3 = ____	7. 2 x 6 = ____
3. 2 x 7 = ____	8. 2 x 5 = ____
4. 2 x 9 = ____	9. 2 x 10= ____
5. 2 x 2 = ____	10. 2 x 1 = ____

Complete this quiz without any help.

1. $2 \times 8 =$ ____
2. $2 \times 2 =$ ____
3. $2 \times 1 =$ ____
4. $2 \times 10 =$ ____
5. $2 \times 5 =$ ____
6. $2 \times 9 =$ ____
7. $2 \times 7 =$ ____
8. $2 \times 3 =$ ____
9. $2 \times 6 =$ ____
10. $2 \times 4 =$ ____

STUDENT WORKSHEETS: CHAPTER 4

LEARNING THE THREE'S

Chapter 4: Learning the Song

The "Three's Song"
(To the tune of "Jingle Bells")

Three, Six, Nine,
Twelve, Fifteen,
Eighteen, Twenty-one,
Twenty-four, Twenty-seven,
Don't forget old Thirty.

3, 6, 9, 12, 15, 18,21, 24, 27,
Don't forget old 30.

Chapter 4: Numbering the Fingers

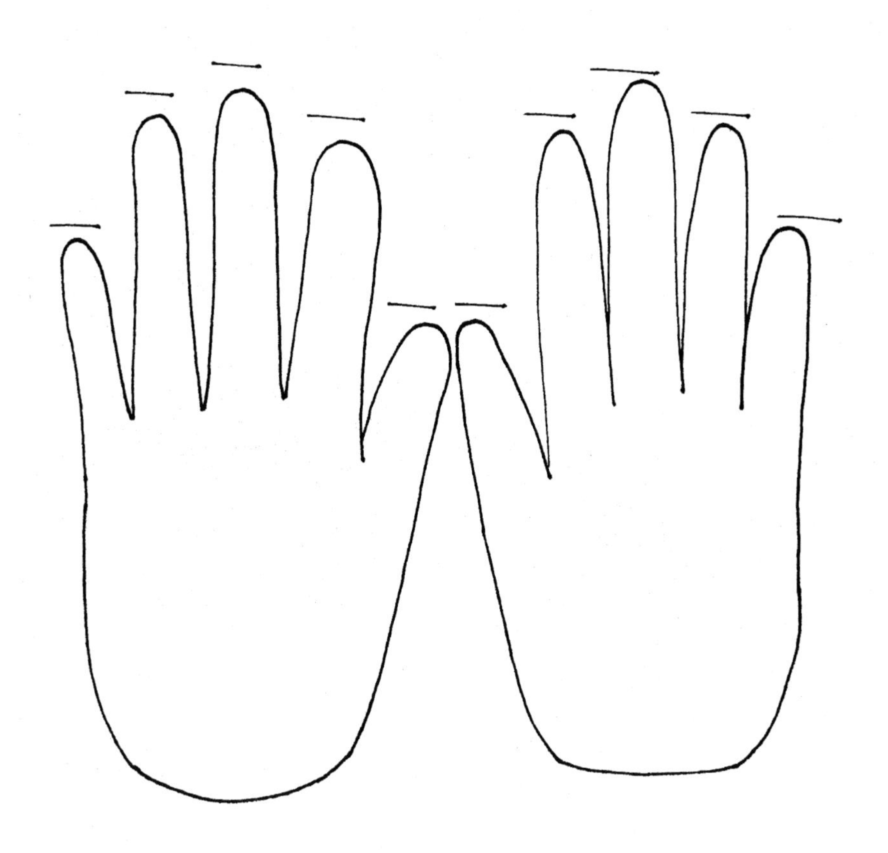

Color the fingers.

2- yellow	9- green	6- black	1- brown	10- gray
5- blue	8- orange	3- red	7- purple	4- pink

Chapter 4: Learning to Multiply

Example A: $3 \times 5 =$ ____

The first number is 3. (Sing the "Three's Song.")
The second number is 5. (Stop on the fifth finger.)
The answer is 15. (The word of the song you're singing when you stop.)

Example B: $3 \times 9 =$ ____

The first number is 3. (Sing the "Three's Song.")
The second number is 9. (Stop on the ninth finger.)
The answer is 27. (The word of the song you're singing when you stop.)

Example C: $3 \times 7 =$ ____

The first number is 3. (Sing the "Three's Song.")
The second number is 7. (Stop on the seventh finger.)
The answer is 21. (The word of the song you're singing when you stop.)

Chapter 4: Practice Page

Draw the picture clue that goes with the "Three's Song."
Write the "Three's Song" in the blanks. ___ ___ ___ ___ ___ ___ ___ ___ ___ ___

1. 3 x 4 = _____	1. 3 x 7 = _____
2. 3 x 9 = _____	2. 3 x 2 = _____
3. 3 x 7 = _____	3. 3 x 6 = _____
4. 3 x 8 = _____	4. 3 x 10 = _____
5. 3 x 5 = _____	5. 3 x 1 = _____
6. 3 x 6 = _____	6. 3 x 9 = _____
7. 3 x 1 = _____	7. 3 x 8 = _____
8. 3 x 10 = _____	8. 3 x 4 = _____
9. 3 x 3 = _____	9. 3 x 5 = _____
10. 3 x 2 = _____	10. 3 x 3 = _____

Chapter 4: Practice Page

1. 3 x 8 = ______	11. 3 x 1 = ______
2. 3 x 9 = ______	12. 3 x 5 = ______
3. 3 x 3 = ______	13. 3 x 4 = ______
4. 3 x 6 = ______	14. 3 x 8 = ______
5. 3 x 1 = ______	15. 3 x 6 = ______
6. 3 x 10= ______	16. 3 x 9 = ______
7. 3 x 5 = ______	17. 3 x 7 = ______
8. 3 x 4 = ______	18. 3 x 3 = ______
9. 3 x 7 = ______	19. 3 x 2 = ______
10. 3 x 2= ______	20. 3 x 10= ______

Write the "Three's Song" in the blanks.

___ ___ ___ ___ ___ ___ ___ ___ ___ ___

Chapter 4: Count by Cards

3	6
9	12
15	18
21	24
27	30

Chapter 4: Multiplication Cards

3 x 1	3 x 2
3 x 3	3 x 4
3 x 5	3 x 6
3 x 7	3 x 8
3 x 9	3x10

Chapter 4: Practice Quiz

Quiz A

1. 3 x 2 = ____	6. 3 x 5 = ____
2. 3 x 4 = ____	7. 3 x 10= ____
3. 3 x 1 = ____	8. 3 x 7 = ____
4. 3 x 9 = ____	9. 3 x 3 = ____
5. 3 x 6 = ____	10. 3 x 8 = ____

Quiz B

1. 2 x 5 = ____	6. 2 x 7 = ____
2. 2 x 1 = ____	7. 2 x 2 = ____
3. 2 x 6 = ____	8. 2 x 3 = ____
4. 2 x 10= ____	9. 2 x 9 = ____
5. 2 x 8 = ____	10. 2 x 4 = ____

Quiz C

1. 2 x 4 = ____	6. 3 x 8 = ____
2. 3 x 3 = ____	7. 2 x 6 = ____
3. 2 x 7 = ____	8. 3 x 5 = ____
4. 3 x 9 = ____	9. 2 x 10= ____
5. 2 x 2 = ____	10. 3 x 1 = ____

Chapter 4: Chapter Quiz

Complete this quiz without any help.

1. $3 \times 8 =$ ____
2. $3 \times 2 =$ ____
3. $3 \times 1 =$ ____
4. $3 \times 10 =$ ____
5. $3 \times 5 =$ ____
6. $3 \times 9 =$ ____
7. $3 \times 7 =$ ____
8. $3 \times 3 =$ ____
9. $3 \times 6 =$ ____
10. $3 \times 4 =$ ____

STUDENT WORKSHEETS: CHAPTER 5

LEARNING THE FOUR'S

The "Four's Song"
(To the tune of "Clementine")

Four, Eight, Twelve,
Sixteen, Twenty,
Twenty-four, Twenty-eight,
Thirty-two, Thirty-six,
Finally, Forty.

4, 8, 12, 16, 20, 24, 28, 32, 36,
Finally, 40.

Chapter 5: Numbering the Fingers

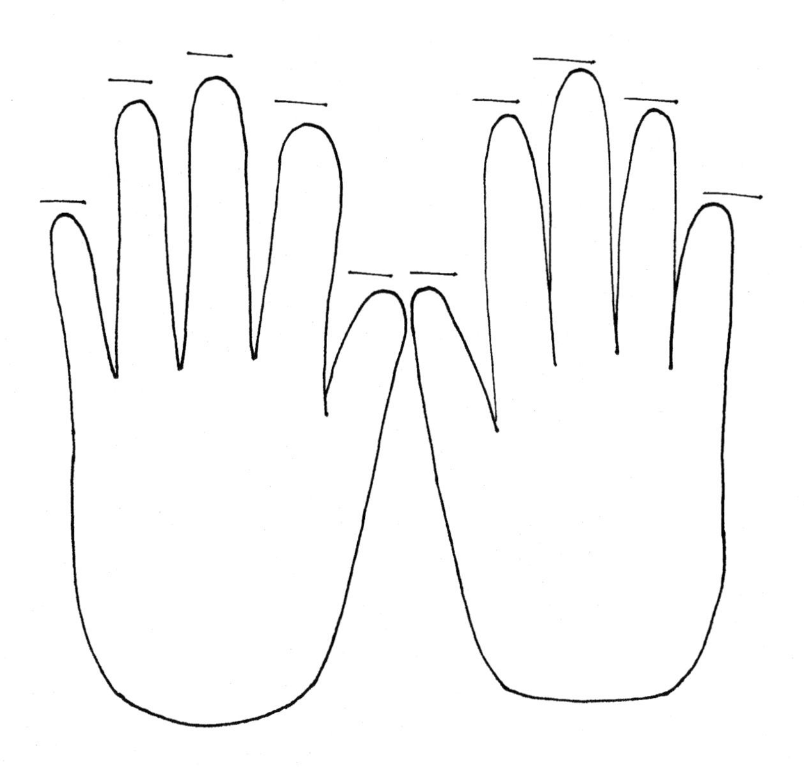

Color the fingers.

1- gray	6- blue	2- purple	7- pink	3- red
8- green	4- yellow	9- brown	5- black	10- orange

Chapter 5: Learning to Multiply

Example A: 4 x 8 = ____

The first number is 4. (Sing the "Four's Song.")
The second number is 8. (Stop on the eighth finger.)
The answer is 32. (The word of the song you're singing when you stop.)

Example B: 4 x 9 = ____

The first number is 4. (Sing the "Four's Song.")
The second number is 9. (Stop on the ninth finger.)
The answer is 36. (The word of the song you're singing when you stop.)

Example C: 4 x 2 = ____

The first number is 4. (Sing the "Four's Song.")
The second number is 2. (Stop on the second finger.)
The answer is 8. (The word of the song you're singing when you stop.)

Chapter 5: Practice Page

Draw the picture clue that goes with the "Four's Song."

Write the "Four's Song" in the blanks.

___ ___ ___ ___ ___ ___ ___ ___ ___ ___

1. 4 x 4 = _____	1. 4 x 7 = _____
2. 4 x 9 = _____	2. 4 x 2 = _____
3. 4 x 7 = _____	3. 4 x 6 = _____
4. 4 x 8 = _____	4. 4 x 10= _____
5. 4 x 5 = _____	5. 4 x 1 = _____
6. 4 x 6 = _____	6. 4 x 9 = _____
7. 4 x 1 = _____	7. 4 x 8 = _____
8. 4 x 10 = _____	8. 4 x 4 = _____
9. 4 x 3 = _____	9. 4 x 5 = _____
10. 4 x 2 = _____	10. 4 x 3 = _____

Chapter 5: Practice Page

1. 4 x 8 = _____	11. 4 x 1 = _____
2. 4 x 9 = _____	12. 4 x 5 = _____
3. 4 x 3 = _____	13. 4 x 4= _____
4. 4 x 6 = _____	14. 4 x 8 = _____
5. 4 x 1 = _____	15. 4 x 6 = _____
6. 4 x 10 = _____	16. 4 x 9 = _____
7. 4 x 5 = _____	17. 4 x 7 = _____
8. 4 x 4 = _____	18. 4 x 3= _____
9. 4 x 7 = _____	19. 4 x 2= _____
10. 4 x 2 = _____	20. 4 x 10=_____

Write the "Four's Song" in the blanks.

____ ____ ____ ____ ____ ____ ____ ____ ____ ____

Chapter 5: Count by Cards

4	8
12	16
20	24
28	32
36	40

Chapter 5: Multiplication Cards

4 x 1	4 x 2
4 x 3	4 x 4
4 x 5	4 x 6
4 x 7	4 x 8
4 x 9	4x10

Chapter 5: Practice Quiz

Quiz A

1. 4 x 2 = ____	6. 4 x 5 = ____
2. 4 x 4 = ____	7. 4 x 10 = ____
3. 4 x 1 = ____	8. 4 x 7 = ____
4. 4 x 9 = ____	9. 4 x 3 = ____
5. 4 x 6 = ____	10. 4 x 8 = ____

Quiz B

1. 3 x 5 = ____	6. 3 x 7 = ____
2. 3 x 1 = ____	7. 3 x 2 = ____
3. 3 x 6 = ____	8. 3 x 3 = ____
4. 3 x 10 = ____	9. 3 x 9 = ____
5. 3 x 8 = ____	10. 3 x 4 = ____

Quiz C

1. 3 x 4 = ____	6. 4 x 8 = ____
2. 4 x 3 = ____	7. 3 x 6 = ____
3. 3 x 7 = ____	8. 4 x 5 = ____
4. 4 x 9 = ____	9. 3 x 10 = ____
5. 3 x 2 = ____	10. 4 x 1 = ____

Chapter 5: Chapter Quiz

Complete this quiz without any help.

1. $4 \times 8 =$ ____
2. $4 \times 2 =$ ____
3. $4 \times 1 =$ ____
4. $4 \times 10 =$ ____
5. $4 \times 5 =$ ____
6. $4 \times 9 =$ ____
7. $4 \times 7 =$ ____
8. $4 \times 3 =$ ____
9. $4 \times 6 =$ ____
10. $4 \times 4 =$ ____

STUDENT WORKSHEETS: CHAPTER 6

LEARNING THE FIVE'S

Chapter 6: Learning the Song

The "Five's Song"
(To the tune of "Pop Goes the Weasel")

Five, Ten, Fifteen,
Twenty, Twenty-five,
Thirty, Thirty-five, Forty,
Forty-five, and Fifty.

5, 10, 15, 20, 25,
30, 35, 40, 45, and 50.

Chapter 6: Numbering the Fingers

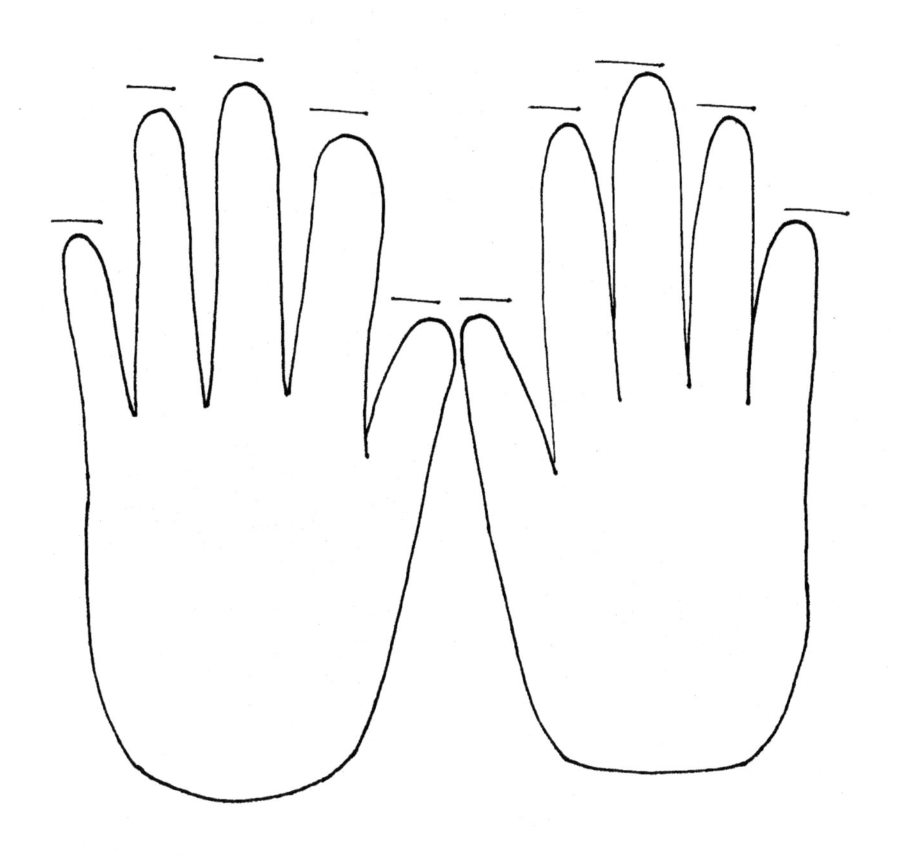

Color the fingers.

4- brown	8- black	5- green	9- gray	10- yellow
6- purple	1- pink	7- orange	2- blue	3- red

Chapter 6: Learning to Multiply

Example A: 5 x 6 = ____

The first number is 5. (Sing the "Five's Song.")
The second number is 6. (Stop on the sixth finger.)
The answer is 30. (The word of the song you're singing when you stop.)

Example B: 5 x 1 = ____

The first number is 5. (Sing the "Five's Song.")
The second number is 1. (Stop on the first finger.)
The answer is 5. (The word of the song you're singing when you stop.)

Example C: 5 x 10 = ____

The first number is 5. (Sing the "Five's Song.")
The second number is 10. (Stop on the tenth finger.)
The answer is 50. (The word of the song you're singing when you stop.)

Chapter 6: Practice Page

Draw the picture clue that goes with the "Five's Song."
Write the "Five's Song" in the blanks. ___ ___ ___ ___ ___ ___ ___ ___ ___ ___

1. 5 x 4 = _____	1. 5 x 7 = _____
2. 5 x 9 = _____	2. 5 x 2 = _____
3. 5 x 7 = _____	3. 5 x 6 = _____
4. 5 x 8 = _____	4. 5 x 10 = _____
5. 5 x 5 = _____	5. 5 x 1 = _____
6. 5 x 6 = _____	6. 5 x 9 = _____
7. 5 x 1 = _____	7. 5 x 8 = _____
8. 5 x 10 = _____	8. 5 x 4 = _____
9. 5 x 3 = _____	9. 5 x 5 = _____
10. 5 x 2 = _____	10. 5 x 3 = _____

Chapter 6: Practice Page

1. 5 x 8 = _____	11. 5 x 1 = _____
2. 5 x 9 = _____	12. 5 x 5 = _____
3. 5 x 3 = _____	13. 5 x 4 = _____
4. 5 x 6 = _____	14. 5 x 8 = _____
5. 5 x 1 = _____	15. 5 x 6 = _____
6. 5 x 10= _____	16. 5 x 9 = _____
7. 5 x 5 = _____	17. 5 x 7 = _____
8. 5 x 4 = _____	18. 5 x 3 = _____
9. 5 x 7 = _____	19. 5 x 2 = _____
10. 5 x 2 = _____	20. 5 x 10=_____

Write the "Five's Song" in the blanks.

___ ___ ___ ___ ___ ___ ___ ___ ___ ___

Chapter 6: Count by Cards

5	10
15	20
25	30
35	40
45	50

Chapter 6: Multiplication Cards

5 x 1	5 x 2
5 x 3	5 x 4
5 x 5	5 x 6
5 x 7	5 x 8
5 x 9	5x10

Chapter 6: Practice Quiz

Quiz A

1. 5 x 2 = ____	6. 5 x 5 = ____
2. 5 x 4 = ____	7. 5 x 10= ____
3. 5 x 1 = ____	8. 5 x 7 = ____
4. 5 x 9 = ____	9. 5 x 3 = ____
5. 5 x 6 = ____	10. 5 x 8 = ____

Quiz B

1. 4 x 5 = ____	6. 3 x 7 = ____
2. 3 x 1 = ____	7. 3 x 2 = ____
3. 4 x 6 = ____	8. 4 x 3 = ____
4. 3 x 10= ____	9. 4 x 9 = ____
5. 4 x 8 = ____	10. 3 x 4 = ____

Quiz C

1. 4 x 4 = ____	6. 5 x 8 = ____
2. 5 x 3 = ____	7. 3 x 6 = ____
3. 3 x 7 = ____	8. 2 x 5 = ____
4. 2 x 9 = ____	9. 4 x 10= ____
5. 4 x 2 = ____	10. 5 x 1 = ____

Chapter 6: Chapter Quiz

Complete this quiz without any help.

1. $5 \times 8 =$ ____
2. $5 \times 2 =$ ____
3. $5 \times 1 =$ ____
4. $5 \times 10 =$ ____
5. $5 \times 5 =$ ____
6. $5 \times 9 =$ ____
7. $5 \times 7 =$ ____
8. $5 \times 3 =$ ____
9. $5 \times 6 =$ ____
10. $5 \times 4 =$ ____

STUDENT WORKSHEETS: CHAPTER 7

LEARNING THE ZERO'S AND ONE'S

Chapter 7: Instructional Page The Zero's Rule

Example A

Example B

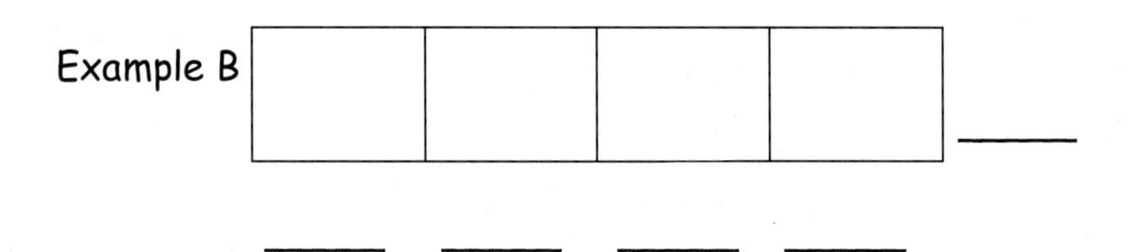

Example C

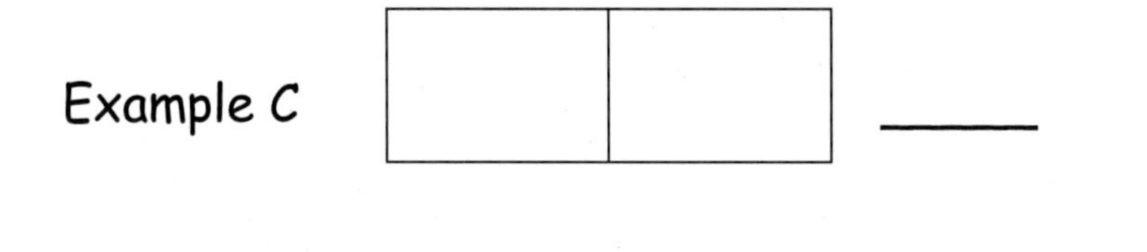

Example D

Example E $0 \times 4 = 0$

Example F $3 \times 0 = 0$

Example G $0 \times 9 = 0$

Example H $6 \times 0 = 0$

Example A

___ ___ ___ ___ ___ ___ ___ ___ ___ ___

Example B

1 x 5 = 5

Example C 1 x 8 = ____

The One's Magic

Example D 1 x 6 = ____

Example E 1 x 9 = ____

Example F 7 x 1 = ____

Example G 1 x 3 = ____

Example H 5 x 1 = ____

The Zero's Rule

1. 0 x 4 = _____	6. 6 x 0 = _____
2. 9 x 0 = _____	7. 0 x 1 = _____
3. 0 x 7 = _____	8. 10 x 0 = _____
4. 8 x 0 = _____	9. 0 x 3 = _____
5. 0 x 5 = _____	10. 2 x 0 = _____

The One's Magic

1. 1 x 7 = _____	6. 9 x 1 = _____
2. 2 x 1 = _____	7. 1 x 8 = _____
3. 1 x 6 = _____	8. 4 x 1 = _____
4. 10 x 1 = _____	9. 1 x 5 = _____
5. 1 x 1 = _____	10. 3 x 1 = _____

Chapter 7: Practice Page Zero's and One's

1. 1 x 8 = _____
2. 9 x 0 = _____
3. 2 x 1 = _____
4. 6 x 0 = _____
5. 7 x 1 = _____
6. 0 x 10 = _____
7. 1 x 5 = _____
8. 2 x 0 = _____
9. 7 x 1 = _____
10. 0 x 2 = _____
11. 8 x 0 = _____
12. 1 x 5 = _____
13. 0 x 4 = _____
14. 8 x 1 = _____
15. 6 x 0 = _____
16. 1 x 9 = _____
17. 0 x 7 = _____
18. 3 x 1= _____
19. 1 x 0= _____
20. 10 x 0 = _____

Chapter 7: Practice Quiz

Quiz A

1.	0 x 2 = ____	6.	5 x 0 = ____
2.	0 x 4 = ____	7.	10 x 0= ____
3.	0 x 1 = ____	8.	7 x 0 = ____
4.	0 x 9 = ____	9.	3 x 0 = ____
5.	0 x 6 = ____	10.	8 x 0 = ____

Quiz B

1.	1 x 5 = ____	6.	7 x 1 = ____
2.	1 x 1 = ____	7.	2 x 1 = ____
3.	1 x 6 = ____	8.	3 x 1 = ____
4.	1 x 10= ____	9.	9 x 1 = ____
5.	1 x 8 = ____	10.	4 x 1 = ____

Quiz C

1.	0 x 4 = ____	6.	1 x 8 = ____
2.	1 x 3 = ____	7.	6 x 0 = ____
3.	7 x 0 = ____	8.	5 x 1 = ____
4.	9 x 1 = ____	9.	0 x 10= ____
5.	0 x 2 = ____	10.	1 x 2 = ____

Chapter 7: Chapter Quiz

Complete this quiz without any help.

1. 0 x 8 = ____
2. 2 x 1 = ____
3. 1 x 1 = ____
4. 10 x 0 = ____
5. 0 x 5 = ____
6. 9 x 1 = ____
7. 1 x 7 = ____
8. 3 x 0 = ____
9. 0 x 6 = ____
10. 4 x 1 = ____

STUDENT WORKSHEETS: CHAPTER 8

LEARNING THE SIX'S

Chapter 8: Learning the Song

The "Six's Song"
(To the tune of "Camptown Races")

Six, Twelve, Eighteen,
Twenty-four, Thirty,
Thirty-six, Forty-two, Forty-eight,
Fifty-four, and Sixty.

6, 12, 18, 24, 30,
36, 42, 48, 54, and 60.

Chapter 8: Numbering the Fingers

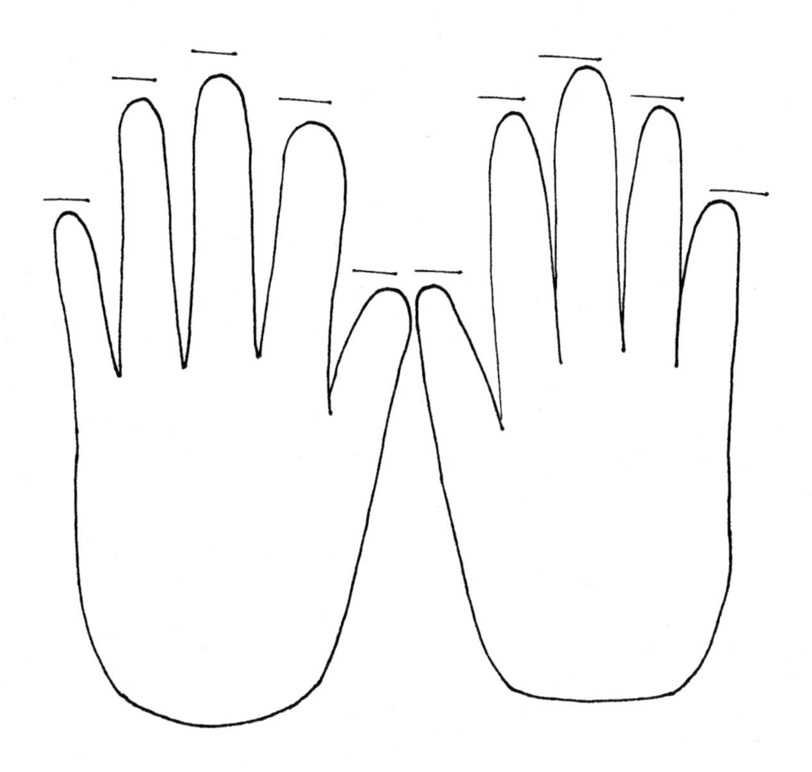

Color the fingers.

10- red	9- blue	7- green	8- gray	4- purple
3- brown	2- black	5- orange	1- pink	6- yellow

Chapter 8: Learning to Multiply

Example A: 6 x 9 = ____

The first number is 6. (Sing the "Six's Song.")
The second number is 9. (Stop on the ninth finger.)
The answer is 54. (The word of the song you're singing when you stop.)

Example B: 6 x 3 = ____

The first number is 6. (Sing the "Six's Song.")
The second number is 3. (Stop on the third finger.)
The answer is 18. (The word of the song you're singing when you stop.)

Example C: 6 x 6 = ____

The first number is 6. (Sing the "Six's Song.")
The second number is 6. (Stop on the sixth finger.)
The answer is 36. (The word of the song you're singing when you stop.)

Chapter 8: Practice Page

Draw the picture clue that goes with the "Six's Song."
Write the "Six's Song" in the blanks. ___ ___ ___ ___ ___ ___ ___ ___ ___ ___

1. 6 x 4 = _____	1. 6 x 7 = _____
2. 6 x 9 = _____	2. 6 x 2 = _____
3. 6 x 7 = _____	3. 6 x 6 = _____
4. 6 x 8 = _____	4. 6 x 10 = _____
5. 6 x 5 = _____	5. 6 x 1 = _____
6. 6 x 6 = _____	6. 6 x 9 = _____
7. 6 x 1 = _____	7. 6 x 8 = _____
8. 6 x 10 = _____	8. 6 x 4 = _____
9. 6 x 3 = _____	9. 6 x 5 = _____
10. 6 x 2 = _____	10. 6 x 3 = _____

Chapter 8: Practice Page

1. 6 x 8 = _____
2. 6 x 9 = _____
3. 6 x 3 = _____
4. 6 x 6 = _____
5. 6 x 1 = _____
6. 6 x 10= _____
7. 6 x 5 = _____
8. 6 x 4 = _____
9. 6 x 7 = _____
10. 6 x 2 = _____
11. 6 x 1 = _____
12. 6 x 5 = _____
13. 6 x 4 = _____
14. 6 x 8 = _____
15. 6 x 6 = _____
16. 6 x 9 = _____
17. 6 x 7 = _____
18. 6 x 3 = _____
19. 6 x 2 = _____
20. 6 x 10=_____

Write the "Six's Song" in the blanks.

___ ___ ___ ___ ___ ___ ___ ___ ___ ___

Chapter 8: Count by Cards

6	12
18	24
30	36
42	48
54	60

6 x 1	6 x 2
6 x 3	6 x 4
6 x 5	6 x 6
6 x 7	6 x 8
6 x 9	6x10

Chapter 8: Practice Quiz

Quiz A

1. 6 x 2 = ____	6. 6 x 5 = ____
2. 6 x 4 = ____	7. 6 x 10= ____
3. 6 x 1 = ____	8. 6 x 7 = ____
4. 6 x 9 = ____	9. 6 x 3 = ____
5. 6 x 6 = ____	10. 6 x 8 = ____

Quiz B

1. 5 x 5 = ____	6. 4 x 7 = ____
2. 4 x 1 = ____	7. 5 x 2 = ____
3. 5 x 6 = ____	8. 4 x 3 = ____
4. 4 x 10= ____	9. 5 x 9 = ____
5. 5 x 8 = ____	10. 4 x 4 = ____

Quiz C

1. 6 x 4 = ____	6. 6 x 8 = ____
2. 5 x 3 = ____	7. 5 x 6 = ____
3. 4 x 7 = ____	8. 4 x 5 = ____
4. 3 x 9 = ____	9. 3 x 10= ____
5. 2 x 2 = ____	10. 2 x 1 = ____

Chapter 8: Chapter Quiz

Complete this quiz without any help.

1. $6 \times 8 =$ ____
2. $6 \times 2 =$ ____
3. $6 \times 1 =$ ____
4. $6 \times 10 =$ ____
5. $6 \times 5 =$ ____
6. $6 \times 9 =$ ____
7. $6 \times 7 =$ ____
8. $6 \times 3 =$ ____
9. $6 \times 6 =$ ____
10. $6 \times 4 =$ ____

STUDENT WORKSHEETS: CHAPTER 9

LEARNING THE SEVEN'S

Chapter 9: Learning the Song

The "Seven's Song"
(To the tune of "Twinkle, Twinkle, Little Star")

Seven, Fourteen, Twenty-one,
Twenty-eight, Thirty-five,
Forty-two, Forty-nine, Fifty-six, Sixty-three,
and Seventy. Can't you see?
Sevens are so fun to do. Don't you think so too?

7, 14, 21, 28, 35, 42, 49, 56, 63, and 70.
Can't you see? Sevens are so fun to do.
Don't you think so too?

Chapter 9: Numbering the Fingers

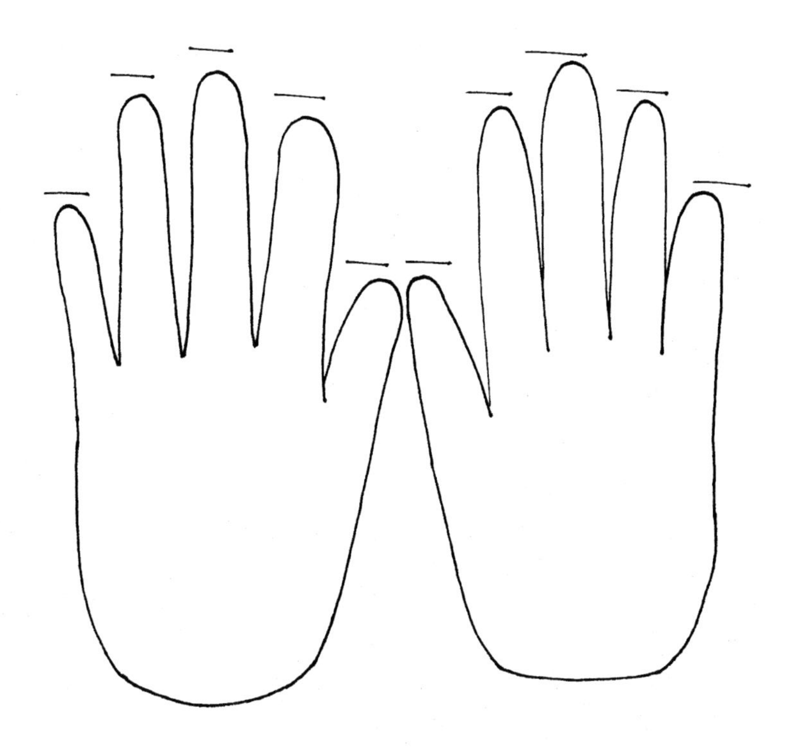

Color the fingers.

1- red	3- orange	2- green	4- brown	6- purple
5- gray	9- black	7- blue	10- yellow	8- pink

Chapter 9: Learning to Multiply

Example A: 7 x 4 = ____

The first number is 7. (Sing the "Seven's Song.")
The second number is 4. (Stop on the fourth finger.)
The answer is 28. (The word of the song you're singing when you stop.)

Example B: 7 x 9 = ____

The first number is 7. (Sing the "Seven's Song.")
The second number is 9. (Stop on the ninth finger.)
The answer is 63. (The word of the song you're singing when you stop.)

Example C: 7 x 3 = ____

The first number is 7. (Sing the "Seven's Song.")
The second number is 3. (Stop on the third finger.)
The answer is 21. (The word of the song you're singing when you stop.)

Chapter 9: Practice Page

Draw the picture clue that goes with the "Seven's Song."
Write the "Seven's Song" in the blanks. ___ ___ ___ ___ ___ ___ ___ ___ ___ ___

1. 7 x 4 = _____	1. 7 x 7 = _____
2. 7 x 9 = _____	2. 7 x 2 = _____
3. 7 x 7 = _____	3. 7 x 6 = _____
4. 7 x 8 = _____	4. 7 x 10 = _____
5. 7 x 5 = _____	5. 7 x 1 = _____
6. 7 x 6 = _____	6. 7 x 9 = _____
7. 7 x 1 = _____	7. 7 x 8 = _____
8. 7 x 10 = _____	8. 7 x 4 = _____
9. 7 x 3 = _____	9. 7 x 5 = _____
10. 7 x 2 = _____	10. 7 x 3 = _____

Chapter 9: Practice Page

1. 7 x 8 = _____
2. 7 x 9 = _____
3. 7 x 3 = _____
4. 7 x 6 = _____
5. 7 x 1 = _____
6. 7 x 10= _____
7. 7 x 5 = _____
8. 7 x 4 = _____
9. 7 x 7 = _____
10. 7 x 2 = _____
11. 7 x 1 = _____
12. 7 x 5 = _____
13. 7 x 4 = _____
14. 7 x 8 = _____
15. 7 x 6 = _____
16. 7 x 9 = _____
17. 7 x 7 = _____
18. 7 x 3 = _____
19. 7 x 2= _____
20. 7 x 10=_____

Write the "Seven's Song" in the blanks.

___ ___ ___ ___ ___ ___ ___ ___ ___ ___

Chapter 9: Count by Cards

7	14
21	28
35	42
49	56
63	70

7 x 1	7 x 2
7 x 3	7 x 4
7 x 5	7 x 6
7 x 7	7 x 8
7 x 9	7x10

Chapter 9: Practice Quiz

Quiz A

1.	7 x 2 = ____	6.	7 x 5 = ____
2.	7 x 4 = ____	7.	7 x 10= ____
3.	7 x 1 = ____	8.	7 x 7 = ____
4.	7 x 9 = ____	9.	7 x 3 = ____
5.	7 x 6 = ____	10.	7 x 8 = ____

Quiz B

1.	6 x 5 = ____	6.	5 x 7 = ____
2.	5 x 1 = ____	7.	6 x 2 = ____
3.	6 x 6 = ____	8.	5 x 3 = ____
4.	5 x 10= ____	9.	6 x 9 = ____
5.	6 x 8= ____	10.	5 x 4 = ____

Quiz C

1.	0 x 4 = ____	6.	1 x 8 = ____
2.	2 x 3 = ____	7.	3 x 6 = ____
3.	4 x 7 = ____	8.	5 x 5 = ____
4.	6 x 9 = ____	9.	7 x 10= ____
5.	5 x 2 = ____	10.	6 x 1 = ____

Chapter 9: Chapter Quiz

Complete this quiz without any help.

1. 7 x 8 = ____

2. 7 x 2 = ____

3. 7 x 1 = ____

4. 7 x 10 = ____

5. 7 x 5 = ____

6. 7 x 9 = ____

7. 7 x 7 = ____

8. 7 x 3 = ____

9. 7 x 6 = ____

10. 7 x 4 = ____

STUDENT WORKSHEETS: CHAPTER 10

LEARNING THE EIGHT'S

The "Eight's Song"
(To the tune of "Mary Had A Little Lamb")
Eight, Sixteen, Twenty-four,
Thirty-two, Forty,
Forty-eight, Fifty-six, Sixty-four,
Seventy-two, and Eighty.

8, 16, 24, 32, 40,
48, 56, 64, 72, and 80.

Chapter 10: Numbering the Fingers

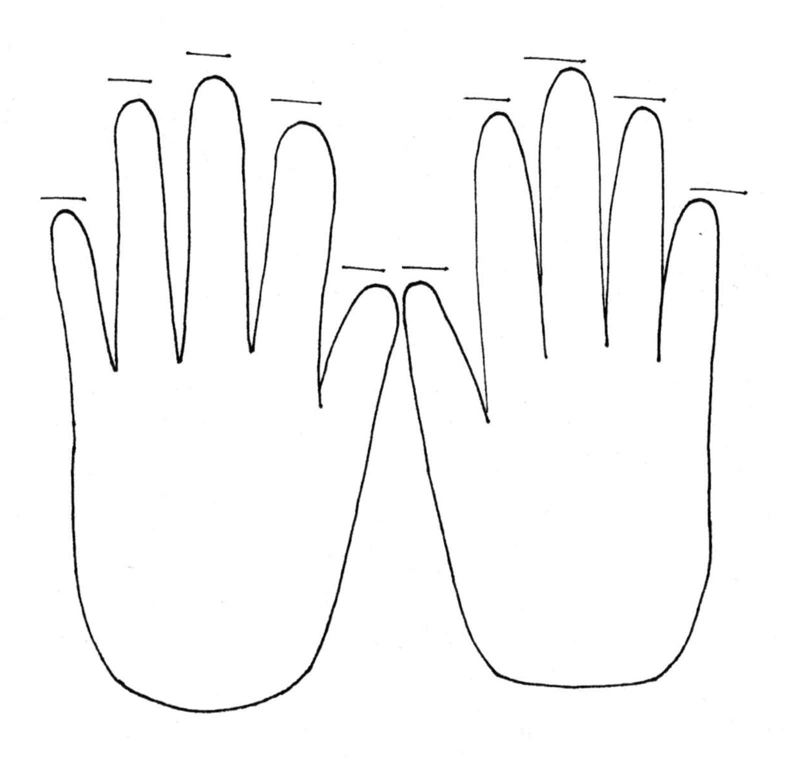

Color the fingers.

1- brown	3- purple	2- pink	4- red	6- orange
5- green	9- gray	7- yellow	10- blue	8- black

Chapter 10: Learning to Multiply

Example A: 8 x 7 = ____

The first number is 8. (Sing the "Eight's Song.")
The second number is 7. (Stop on the seventh finger.)
The answer is 56. (The word of the song you're singing when you stop.)

Example B: 8 x 2 = ____

The first number is 8. (Sing the "Eight's Song.")
The second number is 2. (Stop on the second finger.)
The answer is 16. (The word of the song you're singing when you stop.)

Example C: 8 x 5 = ____

The first number is 8. (Sing the "Eight's Song.")
The second number is 5. (Stop on the fifth finger.)
The answer is 40. (The word of the song you're singing when you stop.)

Chapter 10: Practice Page

Draw the picture clue that goes with the "Eight's Song."	
Write the "Eight's Song" in the blanks. ___ ___ ___ ___ ___ ___ ___ ___ ___ ___	
1. 8 x 4 = _____	1. 8 x 7 = _____
2. 8 x 9 = _____	2. 8 x 2 = _____
3. 8 x 7 = _____	3. 8 x 6 = _____
4. 8 x 8 = _____	4. 8 x 10 = _____
5. 8 x 5 = _____	5. 8 x 1 = _____
6. 8 x 6 = _____	6. 8 x 9 = _____
7. 8 x 1 = _____	7. 8 x 8 = _____
8. 8 x 10 = _____	8. 8 x 4 = _____
9. 8 x 3 = _____	9. 8 x 5 = _____
10. 8 x 2 = _____	10. 8 x 3 = _____

Chapter 10: Practice Page

1. $8 \times 8 =$ ______
2. $8 \times 9 =$ ______
3. $8 \times 3 =$ ______
4. $8 \times 6 =$ ______
5. $8 \times 1 =$ ______
6. $8 \times 10 =$ ______
7. $8 \times 5 =$ ______
8. $8 \times 4 =$ ______
9. $8 \times 7 =$ ______
10. $8 \times 2 =$ ______
11. $8 \times 1 =$ ______
12. $8 \times 5 =$ ______
13. $8 \times 4 =$ ______
14. $8 \times 8 =$ ______
15. $8 \times 6 =$ ______
16. $8 \times 9 =$ ______
17. $8 \times 7 =$ ______
18. $8 \times 3 =$ ______
19. $8 \times 2 =$ ______
20. $8 \times 10 =$ ______

Write the "Eight's Song" in the blanks.

___ ___ ___ ___ ___ ___ ___ ___ ___ ___

Chapter 10: Count by Cards

8	16
24	32
40	48
56	64
72	80

Chapter 10: Multiplication Cards

8 x 1	8 x 2
8 x 3	8 x 4
8 x 5	8 x 6
8 x 7	8 x 8
8 x 9	8x10

Chapter 10: Practice Quiz

Quiz A

1. 8 x 2 = ____	6. 8 x 5 = ____
2. 8 x 4 = ____	7. 8 x 10= ____
3. 8 x 1 = ____	8. 8 x 7 = ____
4. 8 x 9 = ____	9. 8 x 3 = ____
5. 8 x 6 = ____	10. 8 x 8 = ____

Quiz B

1. 7 x 5 = ____	6. 6 x 7 = ____
2. 6 x 1 = ____	7. 7 x 2 = ____
3. 7 x 6 = ____	8. 6 x 3 = ____
4. 6 x 10= ____	9. 7 x 9 = ____
5. 7 x 8 = ____	10. 6 x 4 = ____

Quiz C

1. 8 x 4 = ____	6. 3 x 8 = ____
2. 7 x 3 = ____	7. 2 x 6 = ____
3. 6 x 7 = ____	8. 1 x 5 = ____
4. 5 x 9 = ____	9. 0 x 10= ____
5. 4 x 2 = ____	10. 8 x 1 = ____

Chapter 10: Chapter Quiz

Complete this quiz without any help.

1. 8 x 8 = ____

2. 8 x 2 = ____

3. 8 x 1 = ____

4. 8 x 10 = ____

5. 8 x 5 = ____

6. 8 x 9 = ____

7. 8 x 7 = ____

8. 8 x 3 = ____

9. 8 x 6 = ____

10. 8 x 4 = ____

STUDENT WORKSHEETS: CHAPTER 11

LEARNING THE NINE'S

Chapter 11: Learning the Song

The "Nine's Song"
(To the tune of "For He's A Jolly Good Fellow")

Nine, Eighteen, Twenty-seven,
Thirty-six, Forty-five, Fifty-four,
Sixty-three, Seventy-two, Eighty-one,
And then comes Ninety.

9, 18, 27, 36, 45, 54, 63, 72, 81,
And then comes 90.

Chapter 11: Numbering the Fingers

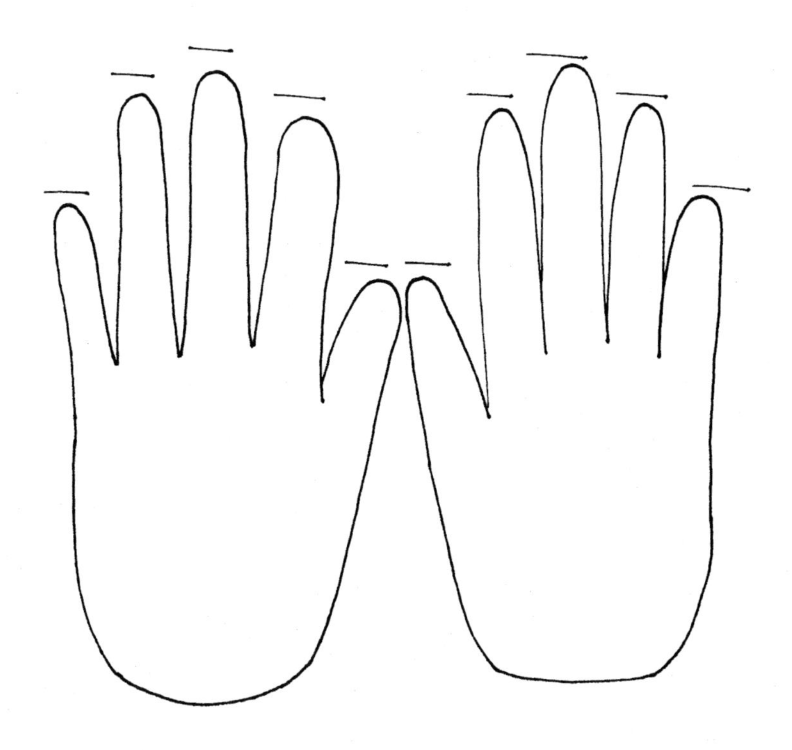

Color the fingers.

1- green	3- gray	2- yellow	4- blue	6- black
5- brown	9- purple	7- pink	10- red	8- orange

Chapter 11: Learning to Multiply

Example A: $9 \times 3 =$ ____

The first number is 9. (Sing the "Nine's Song.")
The second number is 3. (Stop on the third finger.)
The answer is 27. (The word of the song you're singing when you stop.)

Example B: $9 \times 5 =$ ____

The first number is 9. (Sing the "Nine's Song.")
The second number is 5. (Stop on the fifth finger.)
The answer is 45. (The word of the song you're singing when you stop.)

Example C: $9 \times 9 =$ ____

The first number is 9. (Sing the "Nine's Song.")
The second number is 9. (Stop on the ninth finger.)
The answer is 81. (The word of the song you're singing when you stop.)

Chapter 11: Practice Page

Draw the picture clue that goes with the "Nine's Song."

Write the "Nine's Song" in the blanks.

___ ___ ___ ___ ___ ___ ___ ___ ___ ___

1. 9 x 4 = _____	1. 9 x 7 = _____
2. 9 x 9 = _____	2. 9 x 2 = _____
3. 9 x 7 = _____	3. 9 x 6 = _____
4. 9 x 8 = _____	4. 9 x 10 = _____
5. 9 x 5 = _____	5. 9 x 1 = _____
6. 9 x 6 = _____	6. 9 x 9 = _____
7. 9 x 1 = _____	7. 9 x 8 = _____
8. 9 x 10 = _____	8. 9 x 4 = _____
9. 9 x 3 = _____	9. 9 x 5 = _____
10. 9 x 2 = _____	10. 9 x 3 = _____

Chapter 11: Practice Page

1. 9 x 8 = _____	11. 9 x 1 = _____
2. 9 x 9 = _____	12. 9 x 5 = _____
3. 9 x 3 = _____	13. 9 x 4= _____
4. 9 x 6 = _____	14. 9 x 8 = _____
5. 9 x 1 = _____	15. 9 x 6 = _____
6. 9 x 10= _____	16. 9 x 9 = _____
7. 9 x 5 = _____	17. 9 x 7 = _____
8. 9 x 4 = _____	18. 9 x 3 = _____
9. 9 x 7 = _____	19. 9 x 2 = _____
10. 9 x 2 = _____	20. 9 x 10=_____

Write the "Nine's Song" in the blanks.

___ ___ ___ ___ ___ ___ ___ ___ ___ ___

Chapter 11: Count by Cards

9	18
27	36
45	54
63	72
81	90

Chapter 11: Multiplication Cards

9 x 1	9 x 2
9 x 3	9 x 4
9 x 5	9 x 6
9 x 7	9 x 8
9 x 9	9x10

Chapter 11: Practice Quiz

Quiz A

1.	9 x 2 = ____	6.	9 x 5 = ____
2.	9 x 4 = ____	7.	9 x 10= ____
3.	9 x 1 = ____	8.	9 x 7 = ____
4.	9 x 9 = ____	9.	9 x 3 = ____
5.	9 x 6 = ____	10.	9 x 8 = ____

Quiz B

1.	8 x 5 = ____	6.	7 x 7 = ____
2.	7 x 1 = ____	7.	8 x 2 = ____
3.	8 x 6 = ____	8.	7 x 3 = ____
4.	7 x 10= ____	9.	8 x 9 = ____
5.	8 x 8 = ____	10.	7 x 4 = ____

Quiz C

1.	9 x 4 = ____	6.	8 x 7 = ____
2.	7 x 3 = ____	7.	6 x 6 = ____
3.	5 x 7 = ____	8.	4 x 5 = ____
4.	3 x 9 = ____	9.	2 x 10= ____
5.	1 x 2 = ____	10.	0 x 1 = ____

Complete this quiz without any help.

1. 9 x 8 = ____
2. 9 x 2 = ____
3. 9 x 1 = ____
4. 9 x 10 = ____
5. 9 x 5 = ____
6. 9 x 9 = ____
7. 9 x 7 = ____
8. 9 x 3 = ____
9. 9 x 6 = ____
10. 9 x 4 = ____

STUDENT WORKSHEETS: CHAPTER 12

LEARNING THE TEN'S

Chapter 12: Learning the Song

The "Ten's Song"
(To the tune of "Three Blind Mice")

Ten, Twenty, Thirty,
Forty, Fifty, Sixty,
Seventy, Eighty, Ninety,
And One Hundred.
Sing the Tens. That's the End,

10, 20, 30, 40, 50, 60, 70, 80, 90,
And 100. Sing the tens. That's the end.

Chapter 12: Numbering the Fingers

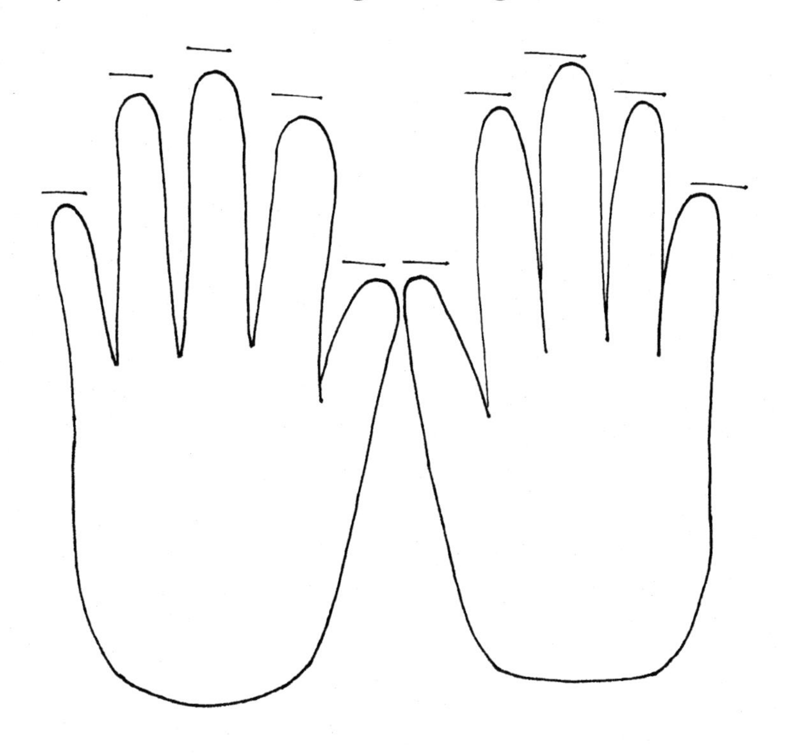

Color the fingers.

1- red	3- orange	2- pink	4- purple	6- brown
5- black	9- blue	7- yellow	10- green	8- gray

Chapter 12: Learning to Multiply

Example A: 10 x 6 = ____

The first number is 10. (Sing the "Ten's Song.")
The second number is 6. (Stop on the sixth finger.)
The answer is 60. (The word of the song you're singing when you stop.)

Example B: 10 x 2 = ____

The first number is 10. (Sing the "Ten's Song.")
The second number is 2. (Stop on the second finger.)
The answer is 20. (The word of the song you're singing when you stop.)

Example C: 10 x 8 = ____

The first number is 10. (Sing the "Ten's Song.")
The second number is 8. (Stop on the eighth finger.)
The answer is 80. (The word of the song you're singing when you stop.)

Chapter 12: Practice Page

Draw the picture clue that goes with the "Ten's Song."	
Write the "Ten's Song" in the blanks. ___ ___ ___ ___ ___ ___ ___ ___ ___ ___	
1. 10 x 4 = _____	1. 10 x 7 = _____
2. 10 x 9 = _____	2. 10 x 2 = _____
3. 10 x 7 = _____	3. 10 x 6 = _____
4. 10 x 8 = _____	4. 10 x 10 = _____
5. 10 x 5 = _____	5. 10 x 1 = _____
6. 10 x 6 = _____	6. 10 x 9 = _____
7. 10 x 1 = _____	7. 10 x 8 = _____
8. 10 x 10 = _____	8. 10 x 4 = _____
9. 10 x 3 = _____	9. 10 x 5 = _____
10. 10 x 2 = _____	10. 10 x 3 = _____

Chapter 12: Practice Page

1. 10 x 8 = _____	11. 10 x 1 = _____
2. 10 x 9 = _____	12. 10 x 5 = _____
3. 10 x 3 = _____	13. 10 x 4 = _____
4. 10 x 6 = _____	14. 10 x 8 = _____
5. 10 x 1 = _____	15. 10 x 6 = _____
6. 10 x 10= _____	16. 10 x 9 = _____
7. 10 x 5 = _____	17. 10 x 7 = _____
8. 10 x 4 = _____	18. 10 x 3 = _____
9. 10 x 7 = _____	19. 10 x 2 = _____
10. 10 x 2 = _____	20. 10 x 10=_____

Write the "Ten's Song" in the blanks.

___ ___ ___ ___ ___ ___ ___ ___ ___ ___

Chapter 12: Count by Cards

10	20
30	40
50	60
70	80
90	100

Chapter 12: Multiplication Cards

10x 1	10x 2
10x 3	10x 4
10x 5	10x 6
10x 7	10x 8
10x 9	10x10

Chapter 12: Practice Quiz

Quiz A

1. 10 x 2 = ____	6. 10 x 5 = ____
2. 10 x 4 = ____	7. 10 x 10= ____
3. 10 x 1 = ____	8. 10 x 7 = ____
4. 10 x 9 = ____	9. 10 x 3 = ____
5. 10 x 6 = ____	10. 10 x 8 = ____

Quiz B

1. 9 x 5 = ____	6. 8 x 7 = ____
2. 8 x 1 = ____	7. 9 x 2 = ____
3. 9 x 6 = ____	8. 8 x 3 = ____
4. 8 x 10= ____	9. 9 x 9 = ____
5. 9 x 8 = ____	10. 8 x 4 = ____

Quiz C

1. 0 x 4 = ____	6. 5 x 8 = ____
2. 1 x 3 = ____	7. 6 x 6 = ____
3. 2 x 7 = ____	8. 7 x 5 = ____
4. 3 x 9 = ____	9. 8 x 10= ____
5. 4 x 2 = ____	10. 9 x 1 = ____

Chapter 12: Chapter Quiz

Complete this quiz without any help.

1. $10 \times 8 =$ ____
2. $10 \times 2 =$ ____
3. $10 \times 1 =$ ____
4. $10 \times 10 =$ ____
5. $10 \times 5 =$ ____
6. $10 \times 9 =$ ____
7. $10 \times 7 =$ ____
8. $10 \times 3 =$ ____
9. $10 \times 6 =$ ____
10. $10 \times 4 =$ ____

STUDENT WORKSHEETS: CHAPTER 13

LEARNING THE ELEVEN'S

Chapter 13: Instructional Page

The Eleven's Trick

Example A

___ ___ ___ ___ ___ ___ ___ ___ ___ ___

Example B

11 x 5 = 55

Example C 11 x 8 = ____

Example D 11 x 6 = ____

Example E 11 x 9 = ____

Example F 7 x 11 = ____

Example G 11x 3 = ____

Example H 5 x 11 = ____

Chapter 13: Practice Page

1. 11 x 4 = ______
2. 9 x 11 = ______
3. 11 x 7 = ______
4. 8 x 11 = ______
5. 11 x 5 = ______
6. 6 x 11 = ______
7. 11 x 1 = ______
8. 11 x 0 = ______
9. 11 x 3 = ______
10. 2 x 11 = ______

1. 11 x 7 = ______
2. 2 x 11 = ______
2. 11 x 6 = ______
3. 10 x 11 = ______
4. 11 x 1 = ______
5. 9 x 11 = ______
6. 11 x 8 = ______
7. 4 x 11 = ______
8. 11 x 5 = ______
10. 3 x 11 = ______

Chapter 13: Practice Page

1. 11 x 8 = _____
2. 9 x 11 = _____
3. 2 x 11 = _____
4. 6 x 11 = _____
5. 7 x 11 = _____
6. 11 x 10= _____
7. 11 x 5 = _____
8. 2 x 11 = _____
9. 7 x 11 = _____
10. 11 x 2= _____
11. 8 x 11 = _____
12. 11 x 5 = _____
13. 11 x 4 = _____
14. 8 x 11 = _____
15. 6 x 11 = _____
16. 11 x 9 = _____
17. 11 x 7 = _____
18. 3 x 11 = _____
19. 1 x 11 = _____
20. 10 x 11= _____

Chapter 13: Practice Quiz

Quiz A

1.	11 x 2 = ____	6.	11 x 5 = ____
2.	11 x 4 = ____	7.	11 x 10= ____
3.	11 x 1 = ____	8.	11 x 7 = ____
4.	11 x 9 = ____	9.	11 x 3 = ____
5.	11 x 6 = ____	10.	11 x 8 = ____

Quiz B

1.	9 x 5 = ____	6.	10 x 2 = ____
2.	10 x 1 = ____	7.	9 x 5 = ____
3.	9 x 6 = ____	8.	10 x 3 = ____
4.	10 x 4= ____	9.	9 x 7 = ____
5.	9 x 8 = ____	10.	10 x 10 = ____

Quiz C

1.	0 x 4 = ____	6.	11 x 8 = ____
2.	1 x 3 = ____	7.	6 x 10 = ____
3.	7 x 7 = ____	8.	5 x 6 = ____
4.	9 x 5 = ____	9.	0 x 9= ____
5.	8 x 2 = ____	10.	2 x 2 = ____

Chapter 13: Chapter Quiz

Complete this quiz without any help.

1. $11 \times 8 =$ ____
2. $2 \times 11 =$ ____
3. $11 \times 1 =$ ____
4. $10 \times 11 =$ ____
5. $11 \times 5 =$ ____
6. $9 \times 11 =$ ____
7. $11 \times 7 =$ ____
8. $3 \times 11 =$ ____
9. $11 \times 6 =$ ____
10. $4 \times 11 =$ ____

STUDENT WORKSHEETS: CHAPTER 14

LEARNING THE TWELVE'S

Chapter 14: Learning the Song

The "Twelve's Song"
(To the tune of "Billy Boy")

Twelve, Twenty-four,
Thirty-six, Forty-eight,
Sixty, and Seventy-two,
Eighty-four, Ninety-six,
Then it's One-O-Eight,
And last comes One Hundred Twenty.

12, 24, 36, 48, 60, and 72,
84, 96, Then it's 1-0-8,
And last comes 120.

Chapter 14: Numbering the Fingers

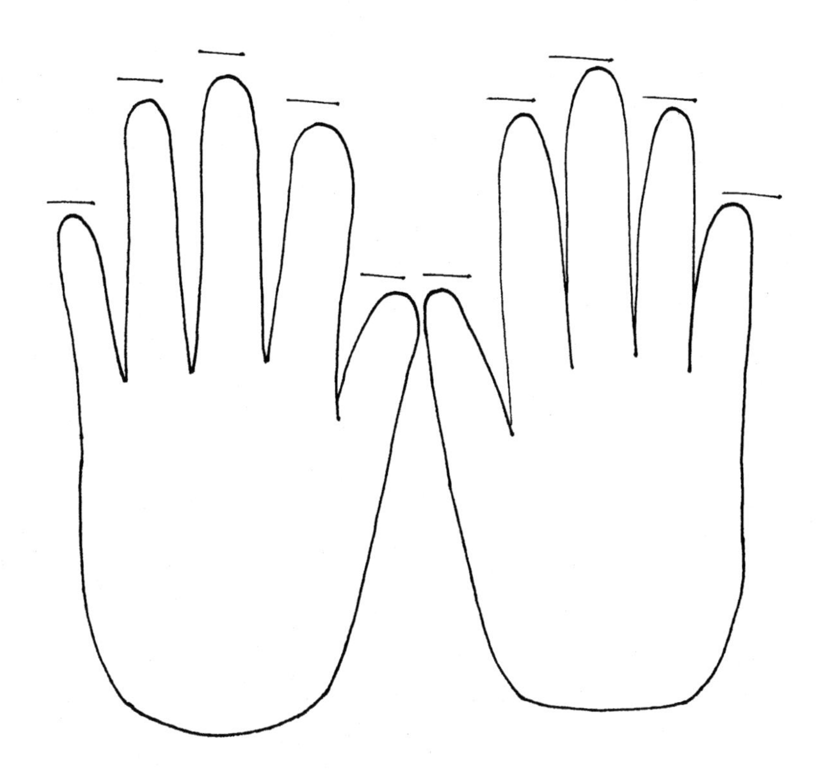

Color the fingers.

1- red	3- orange	2- pink	4- purple	6- brown
5- black	9- blue	7- yellow	10- green	8- gray

Chapter 14: Learning to Multiply

Example A: 12 x 6 = ____

The first number is 12. (Sing the "Twelve's Song.")
The second number is 6. (Stop on the sixth finger.)
The answer is 72. (The word of the song you're singing when you stop.)

Example B: 12 x 2 = ____

The first number is 12. (Sing the "Twelve's Song.")
The second number is 2. (Stop on the second finger.)
The answer is 24. (The word of the song you're singing when you stop.)

Example C: 12 x 8 = ____

The first number is 12. (Sing the "Twelve's Song.")
The second number is 8. (Stop on the eighth finger.)
The answer is 96. (The word of the song you're singing when you stop.)

Chapter 14: Practice Page

Draw the picture clue that goes with the "Twelve's Song."	
Write the "Twelve's Song" in the blanks. ___ ___ ___ ___ ___ ___ ___ ___ ___ ___	
1. 12 x 4 = _____	1. 12 x 7 = _____
2. 12 x 9 = _____	2. 12 x 2 = _____
3. 12 x 7 = _____	3. 12 x 6 = _____
4. 12 x 8 = _____	4. 12 x 10 = _____
5. 12 x 5 = _____	5. 12 x 1 = _____
6. 12 x 6 = _____	6. 12 x 9 = _____
7. 12 x 1 = _____	7. 12 x 8 = _____
8. 12 x 10 = _____	8. 12 x 4 = _____
9. 12 x 3 = _____	9. 12 x 5 = _____
10. 12 x 2 = _____	10. 12 x 3 = _____

Chapter 14: Practice Page

1. 12 x 8 = _____	11. 12 x 1 = _____
2. 12 x 9 = _____	12. 12 x 5 = _____
3. 12 x 3 = _____	13. 12 x 4 = _____
4. 12 x 6 = _____	14. 12 x 8 = _____
5. 12 x 1 = _____	15. 12 x 6 = _____
6. 12 x 10= _____	16. 12 x 9 = _____
7. 12 x 5 = _____	17. 12 x 7 = _____
8. 12 x 4 = _____	18. 12 x 3 = _____
9. 12 x 7 = _____	19. 12 x 2 = _____
10. 12 x 2 = _____	20. 12 x 10=_____

Write the "Twelve's Song" in the blanks.

___ ___ ___ ___ ___ ___ ___ ___ ___ ___

Chapter 14: Count by Cards

12	24
36	48
60	72
84	96
108	120

Chapter 14: Multiplication Cards

12x 1	12x 2
12x 3	12x 4
12x 5	12x 6
12x 7	12x 8
12x 9	12x10

Chapter 14: Practice Quiz

Quiz A

1.	12 x 2 = ____	6.	12 x 5 = ____
2.	12 x 4 = ____	7.	12 x 10= ____
3.	12 x 1 = ____	8.	12 x 7 = ____
4.	12 x 9 = ____	9.	12 x 3 = ____
5.	12 x 6 = ____	10.	12 x 8 = ____

Quiz B

1.	10 x 5 = ____	6.	11 x 7 = ____
2.	11 x 1 = ____	7.	10 x 2 = ____
3.	10 x 6 = ____	8.	11 x 3 = ____
4.	11 x 10= ____	9.	10 x 9 = ____
5.	10 x 8 = ____	10.	11 x 4 = ____

Quiz C

1.	0 x 4 = ____	6.	5 x 8 = ____
2.	11 x 3 = ____	7.	6 x 6 = ____
3.	2 x 7 = ____	8.	7 x 5 = ____
4.	10 x 9 = ____	9.	12 x 10= ____
5.	4 x 2 = ____	10.	9 x 1 = ____

Chapter 14: Chapter Quiz

Complete this quiz without any help.

1. 12 x 8 = ____

2. 12 x 2 = ____

3. 12 x 1 = ____

4. 12 x 10 = ____

5. 12 x 5 = ____

6. 12 x 9 = ____

7. 12 x 7 = ____

8. 12 x 3 = ____

9. 12 x 6 = ____

10. 12 x 4 = ____

STUDENT WORKSHEETS:

EXTRA PRACTICE

Chapter 15: Extra Practice Page

1. 10 x 4 = ____	11. 2 x 8 = ____	21. 10 x 3 = ____
2. 9 x 4 = ____	12. 9 x 2 = ____	22. 1 x 5 = ____
3. 11 x 3 = ____	13. 10 x 5 = ____	23. 8 x 3 = ____
4. 5 x 9 = ____	14. 3 x 7 = ____	24. 12 x 8 = ____
5. 8 x 1 = ____	15. 4 x 6 = ____	25. 6 x 8 = ____
6. 3 x 3 = ____	16. 8 x 1 = ____	26. 6 x 3 = ____
7. 1 x 9 = ____	17. 8 x 5 = ____	27.7 x 10 = ____
8. 6 x 7 = ____	18. 2 x 6 = ____	28. 2 x 4 = ____
9. 9 x 9 = ____	19. 4 x 9 = ____	29. 5 x 8 = ____
10. 7 x 2 = ____	20. 7 x 10= ____	30. 2 x 4 = ____

Chapter 15: Extra Practice Page

1. 10 x 7 = ____	11. 2 x 6 = ____	21. 4 x 3 = ____
2. 7 x 4 = ____	12. 9 x 4 = ____	22.12 x 7 = ____
3. 4 x 6 = ____	13. 6 x 5 = ____	23. 8 x 5 = ____
4. 11 x 9 = ____	14.11 x 7 = ____	24. 4 x 8 = ____
5. 8 x 7 = ____	15. 0 x 6 = ____	25. 6 x 4 = ____
6. 5 x 3 = ____	16. 8 x 9 = ____	26. 7 x 3 = ____
7. 3 x 2 = ____	17. 5 x 5 = ____	27. 9 x 7 = ____
8. 10 x 7 = ____	18. 2 x 4 = ____	28. 4 x 4 = ____
9. 9 x 0 = ____	19. 11 x 9= ____	29. 1 x 8 = ____
10. 5 x 2 = ____	20. 7 x 5 = ____	30.12 x 5 = ____

Chapter 15: Extra Practice Page

1. 10 x 3 = ____	11. 5 x 7 = ____	21. 8 x 3 = ____
2. 3 x 4 = ____	12. 9 x 6=____	22. 1 x 3 = ____
3. 4 x 2 = ____	13. 2 x 5 = ____	23. 8 x 8 = ____
4. 7 x 9 = ____	14.12 x 8 = ____	24. 5 x 8 =____
5. 8 x 6 = ____	15. 3 x 6 = ____	25. 6 x 4 =____
6. 10 x 3 = ____	16.11 x 2 = ____	26. 1 x 3 = ____
7. 3 x 9 = ____	17. 6 x 5 = ____	27. 7 x 7 = ____
8. 8 x 7 = ____	18.2 x 10 = ____	28.10 x 4 = ____
9. 9 x 2 = ____	19. 8 x 9 =____	29. 7 x 3 = ____
10. 7 x 2 = ____	20. 7 x 1=____	30. 5 x 4 = ____

Chapter 15: Extra Practice Page

1. 10 x 6 =____	11.2 x 10 = ____	21. 6 x 2 = ____
2. 6 x 4 = ____	12. 9 x 1 =____	22. 1 x 0 = ____
3. 4 x 8 = ____	13. 7 x 5 = ____	23. 8 x 3 = ____
4. 6 x 9 = ____	14. 9 x 7 = ____	24. 12 x 8=____
5. 8 x 10 = ____	15. 8 x 6 = ____	25. 6 x 12 =____
6. 7 x 3 = ____	16. 8 x 5 = ____	26. 4 x 3 = ____
7. 3 x 6 = ____	17. 5 x 5 = ____	27. 7 x 7 = ____
8. 0 x 7 = ____	18. 2 x 8 =____	28. 5 x 4 = ____
9. 9 x 8 = ____	19. 3 x 9 =____	29. 6 x 6 =____
10. 3 x 2 = ____	20. 7 x 6 = ____	30. 2 x 4 = ____

APPENDIX:

SCORE SHEET AND AWARD

Score Sheet

	Quiz Grade	10	20	30	40	50	60	70	80	90	100
Two's (Ch. 3)											
Three's (Ch. 4)											
Four's (Ch. 5)											
Five's (Ch. 6)											
One's and Zero's (Ch. 7)											
Sixes (Ch. 8)											
Seven's (Ch. 9)											
Eight's (Ch. 10)											
Nine's (Ch. 11)											
Ten's (Ch. 12)											
Eleven's (Ch. 13)											
Twelve's (Ch. 14)											

This certificate is awarded to

____

on this day, ______________

____

for learning how to multiply by

____________________________.

Signed by________________

Instructor

ANSWER KEY

Chapter 3

PAGE 125 PRACTICE PAGE

1. 8
2. 18
3. 14
4. 16
5. 10
6. 12
7. 2
8. 20
9. 6
10. 4
1. 14
2. 4
3. 12
4. 20
5. 2
6. 18
7. 16
8. 8
9. 10
10. 6

PAGE 126 PRACTICE PAGE

1. 16
2. 18
3. 6
4. 12
5. 2
6. 20
7. 10
8. 8
9. 14
10. 4
11. 2
12. 10
13. 8
14. 16
15. 12
16. 18
17. 14
18. 6
19. 4
20. 20

PAGE 129 QUIZ A

1. 4
2. 8
3. 2
4. 18
5. 12
6. 10
7. 20
8. 14
9. 6
10. 16

PAGE 129 QUIZ B

1. 10
2. 2
3. 12
4. 20
5. 16
6. 14
7. 4
8. 6
9. 18
10. 8

PAGE 129 QUIZ C

1. 8
2. 6
3. 14
4. 18
5. 4
6. 16
7. 12
8. 10
9. 20
10. 2

PAGE 130 QUIZ

1. 16
2. 4
3. 2
4. 20
5. 10
6. 18
7. 14
8. 6
9. 12
10. 8

Chapter 4

PAGE 135 PRACTICE PAGE

1. 12
2. 27
3. 21
4. 24
5. 15
6. 18
7. 3
8. 30
9. 9
10. 6
1. 21
2. 6
3. 18
4. 30
5. 3
6. 27
7. 24
8. 12
9. 15
10. 9

PAGE 136 PRACTICE PAGE

1. 24
2. 27
3. 9
4. 18
5. 3
6. 30
7. 15
8. 12
9. 21
10. 6
11. 3
12. 15
13. 12
14. 24
15. 18
16. 27
17. 21
18. 9
19. 6
20. 30

PAGE 139 QUIZ A

1. 6
2. 12
3. 3
4. 27
5. 18
6. 15
7. 30
8. 21
9. 9
10. 24

PAGE 139 QUIZ B

1. 10
2. 2
3. 12
4. 20
5. 16
6. 14
7. 4
8. 6
9. 18
10. 8

PAGE 139 QUIZ C

1. 8
2. 9
3. 14
4. 27
5. 4
6. 24
7. 12
8. 15
9. 20
10. 3

PAGE 140 QUIZ

1. 24
2. 6
3. 3
4. 30
5. 15
6. 27
7. 21
8. 9
9. 18
10. 12

Chapter 5

PAGE 145 PRACTICE PAGE

1. 16
2. 36
3. 28
4. 32
5. 20
6. 24
7. 4
8. 40
9. 12
10. 8

1. 28
2. 8
3. 24
4. 40
5. 4
6. 36
7. 32

8. 16
9. 20
10. 12

8. 28
9. 12
10. 32

PAGE 146 PRACTICE PAGE

1. 32
2. 36
3. 12
4. 24
5. 4
6. 40
7. 20
8. 16
9. 28
10. 8
11. 4
12. 20
13. 16
14. 32
15. 24
16. 36
17. 28
18. 12
19. 8
20. 40

PAGE 149 QUIZ A

1. 8
2. 16
3. 4
4. 36
5. 24
6. 20
7. 40

PAGE 149 QUIZ B

1. 15
2. 3
3. 18
4. 30
5. 24
6. 21
7. 6
8. 9
9. 27
10. 12

PAGE 149 QUIZ C

1. 12
2. 12
3. 21
4. 36
5. 6
6. 32
7. 18
8. 20
9. 30
10. 4

PAGE 150 QUIZ

1. 32
2. 8
3. 4
4. 40
5. 20
6. 36
7. 28
8. 12
9. 24
10. 16

Chapter 6

PAGE 155 PRACTICE PAGE

1. 20
2. 45
3. 35
4. 40
5. 25
6. 30
7. 5
8. 50
9. 15
10. 10
1. 35
2. 10
3. 30
4. 50
5. 5
6. 45
7. 40
8. 20
9. 25
10. 15

PAGE 156 PRACTICE PAGE

1. 40
2. 45
3. 15
4. 30
5. 5
6. 50
7. 25
8. 20
9. 35
10. 10
11. 5
12. 25
13. 20
14. 40
15. 30
16. 45
17. 35
18. 15
19. 10
20. 50

PAGE 159 QUIZ A

1. 10
2. 20
3. 5
4. 45
5. 30
6. 25
7. 50
8. 35
9. 15
10. 40

PAGE 159 QUIZ B

1. 20
2. 3
3. 24
4. 30
5. 32
6. 21
7. 6
8. 12
9. 36
10. 12

PAGE 159 QUIZ C

1. 16
2. 15
3. 21
4. 18
5. 8
6. 40
7. 18
8. 10
9. 40
10. 5

PAGE 160 QUIZ

1. 40
2. 10
3. 5
4. 50
5. 25
6. 45
7. 35
8. 15
9. 30
10. 20

Chapter 7

PAGE 164 PRACTICE PAGE

1. 0
2. 0
3. 0
4. 0
5. 0
6. 0
7. 0
8. 0
9. 0
10. 0

1. 7
2. 2
3. 6
4. 10
5. 1
6. 9
7. 8
8. 4
9. 5
10. 3

PAGE 165 PRACTICE PAGE

1. 8
2. 0
3. 2
4. 0
5. 7
6. 0
7. 5
8. 0
9. 7
10. 0
11. 0
12. 5
13. 0
14. 8
15. 0
16. 9
17. 0
18. 3
19. 0
20. 0

PAGE 166 QUIZ A

1. 0
2. 0
3. 0
4. 0
5. 0
6. 0
7. 0
8. 0
9. 0
10. 0

PAGE 166 QUIZ B

1. 5
2. 1
3. 6
4. 10
5. 8
6. 7
7. 2
8. 3
9. 9
10. 4

PAGE 166 QUIZ C

1. 0
2. 3
3. 0
4. 9
5. 0
6. 8
7. 0
8. 5
9. 0
10. 2

PAGE 167 QUIZ

1. 0
2. 2
3. 1
4. 0
5. 0
6. 9
7. 7
8. 0

9. 0
10. 4

Chapter 8

PAGE 173 PRACTICE PAGE

1. 24
2. 54
3. 42
4. 48
5. 30
6. 36
7. 6
8. 60
9. 18
10. 12
1. 42
2. 12
3. 36
4. 60
5. 6
6. 54
7. 48
8. 24
9. 30
10. 18

PAGE 174 PRACTICE PAGE

1. 48
2. 54
3. 18
4. 36
5. 6
6. 60
7. 30
8. 24
9. 42
10. 12
11. 6
12. 30
13. 24
14. 48
15. 36
16. 54
17. 42
18. 18
19. 12
20. 60

PAGE 177 QUIZ A

1. 12
2. 24
3. 6
4. 54
5. 36
6. 30
7. 60
8. 42
9. 18
10. 48

PAGE 177 QUIZ B

1. 25
2. 4
3. 30
4. 40
5. 40
6. 28
7. 10
8. 12
9. 45
10. 16

PAGE 177 QUIZ C

1. 24
2. 15
3. 28
4. 27
5. 4
6. 48
7. 30
8. 20
9. 30
10. 2

PAGE 178 QUIZ

1. 48
2. 12
3. 6
4. 60
5. 30
6. 54
7. 42
8. 18
9. 36
10. 24

Chapter 9

PAGE 183 PRACTICE PAGE

1. 28
2. 63
3. 49
4. 56
5. 35
6. 42
7. 7
8. 70
9. 21
10. 14

1. 49
2. 14
3. 42
4. 70
5. 7
6. 63
7. 56
8. 28
9. 35
10. 21

PAGE 184 PRACTICE PAGE

1. 56
2. 63
3. 21
4. 42
5. 7
6. 70
7. 35
8. 28
9. 49
10. 14
11. 7

12. 35
13. 28
14. 56
15. 42
16. 63
17. 49
18. 21
19. 14
20. 70

PAGE 187 QUIZ A

1. 14
2. 28
3. 7
4. 63
5. 42
6. 35
7. 70
8. 49
9. 21
10. 56

PAGE 187 QUIZ B

1. 30
2. 5
3. 36
4. 50
5. 48
6. 35
7. 12
8. 15
9. 54
10. 20

PAGE 187 QUIZ C

1. 0
2. 6
3. 28
4. 54
5. 10
6. 8
7. 18
8. 25
9. 70
10. 6

PAGE 188 QUIZ

1. 56
2. 14
3. 7
4. 70
5. 35
6. 63
7. 49
8. 21
9. 42
10. 28

Chapter 10

PAGE 193 PRACTICE PAGE

1. 32
2. 72
3. 56
4. 64
5. 40
6. 48
7. 8
8. 80
9. 24
10. 16

1. 56
2. 16
3. 48
4. 80
5. 8
6. 72
7. 64
8. 32
9. 40
10. 24

PAGE 194 PRACTICE PAGE

1. 64
2. 72
3. 24
4. 48
5. 8
6. 80
7. 40
8. 32
9. 56
10. 16
11. 8
12. 40
13. 32
14. 64
15. 48
16. 72
17. 56
18. 24
19. 16
20. 80

PAGE 197 QUIZ A

1. 16
2. 32
3. 8
4. 72
5. 48
6. 40
7. 80
8. 56
9. 24
10. 64

PAGE 197 QUIZ B

1. 35
2. 6
3. 42
4. 60
5. 56
6. 42
7. 14
8. 18
9. 63
10. 24

PAGE 197 QUIZ C

1. 32
2. 21
3. 42
4. 45
5. 8
6. 24
7. 12
8. 5
9. 0
10. 8

PAGE 198 QUIZ

1. 64
2. 16
3. 8
4. 80
5. 40
6. 72
7. 56
8. 24
9. 48
10. 32

Chapter 11

PAGE 203 PRACTICE PAGE

1. 36
2. 81
3. 63
4. 72
5. 45
6. 54
7. 9
8. 90
9. 27
10. 18
1. 63
2. 18
3. 54
4. 90
5. 9
6. 81
7. 72
8. 36
9. 45
10. 27

PAGE 204 PRACTICE PAGE

1. 72
2. 81
3. 27
4. 54
5. 9
6. 90
7. 45
8. 36
9. 63
10. 18
11. 9
12. 45
13. 36
14. 72
15. 54
16. 81
17. 63
18. 27
19. 18
20. 90

PAGE 207 QUIZ A

1. 18
2. 36
3. 9
4. 81
5. 54
6. 45
7. 90
8. 63
9. 27
10. 72

PAGE 207 QUIZ B

1. 40
2. 7
3. 48
4. 70
5. 64
6. 49
7. 16
8. 21
9. 72
10. 28

PAGE 207 QUIZ C

1. 36
2. 21
3. 35
4. 27
5. 2
6. 56
7. 36
8. 20
9. 20
10. 0

PAGE 208 QUIZ

1. 72
2. 18
3. 9
4. 90
5. 45
6. 81
7. 63
8. 27
9. 54
10. 36

Chapter 12

PAGE 213 PRACTICE PAGE

1. 40
2. 90
3. 70
4. 80
5. 50
6. 60
7. 10
8. 100
9. 30
10. 20
1. 70
2. 20
3. 60
4. 100
5. 10
6. 90
7. 80
8. 40
9. 50
10. 30

PAGE 214 PRACTICE PAGE

1. 80
2. 90
3. 30
4. 60
5. 10
6. 100
7. 50
8. 40
9. 70
10. 20
11. 10
12. 50
13. 40
14. 80
15. 60
16. 90
17. 70
18. 30
19. 20
20. 100

PAGE 217 QUIZ A

1. 20
2. 40
3. 10
4. 90
5. 60
6. 50
7. 100
8. 70
9. 30
10. 80

PAGE 217 QUIZ B

1. 45
2. 8
3. 54
4. 80
5. 72
6. 56
7. 18
8. 24
9. 81
10. 32

PAGE 217 QUIZ C

1. 0
2. 3
3. 14
4. 27
5. 8
6. 40
7. 36
8. 35
9. 80
10. 9

PAGE 218 QUIZ

1. 80
2. 20
3. 10
4. 100
5. 50
6. 90
7. 70
8. 30
9. 60
10. 40

Chapter 13

PAGE 221 PRACTICE PAGE

1. 44
2. 99
3. 77
4. 88
5. 55
6. 66
7. 11
8. 0
9. 33
10. 22
1. 77
2. 22
3. 66
4. 110
5. 11
6. 99
7. 88
8. 44
9. 55
10. 33

PAGE 222 PRACTICE PAGE

1. 88
2. 99
3. 22
4. 66
5. 77
6. 110
7. 55
8. 22
9. 77
10. 22
11. 88
12. 55
13. 44
14. 88
15. 66
16. 99
17. 77
18. 33
19. 11
20. 110

PAGE 223 QUIZ A

1. 22
2. 44
3. 11
4. 99
5. 66
6. 55
7. 110
8. 77
9. 33
10. 88

PAGE 223 QUIZ B

1. 45
2. 10
3. 54
4. 40
5. 72
6. 20
7. 45
8. 30
9. 63
10. 100

PAGE 223 QUIZ C

1. 0
2. 3
3. 49
4. 45
5. 16
6. 88
7. 60
8. 30
9. 0
10. 4

PAGE 224 QUIZ

1. 88
2. 22
3. 11
4. 110
5. 55
6. 99
7. 77
8. 33
9. 66
10. 44

Chapter 14

PAGE 229 PRACTICE PAGE

1. 48
2. 108
3. 84
4. 96
5. 60
6. 72
7. 12
8. 120
9. 36
10. 24

1. 84
2. 24
3. 72
4. 120
5. 12
6. 108
7. 96
8. 48
9. 60
10. 36

PAGE 230 PRACTICE PAGE

1. 96
2. 108
3. 36
4. 72
5. 12
6. 120
7. 60
8. 48
9. 84
10. 24
11. 12
12. 60
13. 48
14. 96
15. 72
16. 108
17. 84
18. 36
19. 24
20. 120

PAGE 233 QUIZ A

1. 24
2. 48
3. 12
4. 108
5. 72
6. 60
7. 120
8. 84
9. 36
10. 96

PAGE 233 QUIZ B

1. 50
2. 11
3. 60
4. 110
5. 80
6. 77
7. 20
8. 33
9. 90
10. 44

PAGE 233 QUIZ C

1. 0
2. 33
3. 14
4. 90
5. 8
6. 40
7. 36
8. 35
9. 120
10. 9

PAGE 234 QUIZ

1. 96
2. 24
3. 12
4. 120
5. 60
6. 108
7. 84
8. 36

9. 72
10. 48

Chapter 15

PAGE 236 PRACTICE

1. 40
2. 36
3. 33
4. 45
5. 8
6. 9
7. 9
8. 42
9. 81
10. 14
11. 16
12. 18
13. 50
14. 21
15. 24
16. 8
17. 40
18. 12
19. 36
20. 70
21. 30
22. 5
23. 24
24. 96
25. 48
26. 18
27. 70
28. 8
29. 40
30. 48

PAGE 237 PRACTICE

1. 70
2. 28
3. 24
4. 99
5. 56
6. 15
7. 6
8. 70
9. 0
10. 10
11. 12
12. 36
13. 30
14. 77
15. 0
16. 72
17. 25
18. 8
19. 99
20. 35
21. 12
22. 84
23. 40
24. 32
25. 24
26. 21
27. 63
28. 16
29. 8
30. 60

PAGE 238 PRACTICE

1. 30
2. 12
3. 8
4. 63
5. 48
6. 30
7. 27
8. 56
9. 18
10. 14
11. 35
12. 54
13. 10
14. 96
15. 18
16. 22
17. 30
18. 20
19. 72
20. 7
21. 24
22. 3
23. 64
24. 40
25. 24
26. 3
27. 49
28. 40
29. 21
30. 20

PAGE 239 PRACTICE

1. 60
2. 24
3. 32
4. 54
5. 80
6. 21
7. 18
8. 0
9. 72
10. 6
11. 20
12. 9
13. 35
14. 63
15. 48
16. 40
17. 25
18. 16
19. 27
20. 42
21. 12
22. 0
23. 24
24. 96
25. 72
26. 12
27. 49
28. 20
29. 36
30. 8

listen|imagine|view|experience

AUDIO BOOK DOWNLOAD INCLUDED WITH THIS BOOK!

In your hands you hold a complete digital entertainment package. Besides purchasing the paper version of this book, this book includes a free download of the audio version of this book. Simply use the code listed below when visiting our website. Once downloaded to your computer, you can listen to the book through your computer's speakers, burn it to an audio CD or save the file to your portable music device (such as Apple's popular iPod) and listen on the go!

How to get your free audio book digital download:

1. Visit www.tatepublishing.com and click on the e|LIVE logo on the home page.
2. Enter the following coupon code:
   7052-d0f3-8df4-f992-fbea-bb62-a81a-d8db
3. Download the audio book from your e|LIVE digital locker and begin enjoying your new digital entertainment package today!